Data Analysis and Its Interpretation: Application in Marketing Research

Dewi Indriani Jusuf & Jonathan Sarwono

Copyright © 2018 Dewi Indriani Jusuf & Jonathan Sarwono

All rights reserved.

ISBN: 9781983265129
ISBN-13: 9781983265129
Published by: Amazon.com, Inc. 410 Terry Avenue North Seattle, Washington 98109 US

DEDICATION

This book is dedicated for students whose concern is a marketing research

CONTENTS

	Acknowledgments	i
1	Basic concepts	1
2	Using Mean for a Comparative Analysis	26
3	Principal Component Analysis to Identify Valid Predictors	29
4	Analyzing Relationship Between More Than One Independent and Dependent Variables Using Canonical Correlation	35
5	A Good Model Making Simulation	43
6	How to Analyze the Data Having Different Measurement Scales	58
7	How to Make a Graph to Present Research Results	91
8	A Marketing Research Example Using Linear Regression Considering the Classical Assumption	107
9	Direct Marketing: Procedures Using Consumers' Data Base	191

ACKNOWLEDGMENTS

There are some changes in IBM SPSS 25 relating to ways of calculating the data using certain analysis procedures. As an example, canonical correlation is now made simple in calculation. In the previous versions, to calculate the data using this procedure a user must use IBM SPSS syntax commands.

In this book, the writers will discuss several procedures that are used in marketing research, such as simulation, canonical correlation and linear regression. Furthermore, there will be a comprehensive discussion on applying a linear regression procedure in a marketing research that will be helpful for students who are conducting their researches. The calculation is done using IBM SPSS version 21, 22, 24 and 25. The examples are applied in the area of marketing research.

Any questions relating to this book can be sent to the authors' email at dewijusuf@iwu.ac.id and jsarwono007@gmail.com.

July 2018

Authors

CHAPTER 1

BASIC CONCEPTS

1.1 Introduction

To make it easier for readers to understand the use of SPSS properly and correctly and to obtain maximum results, the reader needs to understand some basic concepts that serve as a theory for underlying in operating SPSS and interpreting the output correctly. The basic concepts are variables, relationships between variables, confidence interval, significance level, data / case definition, one-tailed and two-tailed hypotheses test, degrees of freedom, critical values, parametric and nonparametric statistics.

1.2 Variables

What is a variable? A variable is defined as " something that may vary or differ " (Brown, 1998: 7). Another more detailed definition says that the variable " is simply a symbol or a concept that can assume any one of a set of values " (Davis, 1998: 23). Cramer and Howitt (2006) define variables as: " a characteristic that consists of two or more categories or values "

The first definition states that the variable is something different or varied, the emphasis of the word something is clarified in the second definition of the symbol or concept that is assumed as a set of values. While the third definition describes a variable as a characteristic that has two or more categories or values. The notion provides an explanation that the variable is something either abstract or concrete that has at least two or more categories or has a value. Category refers to a variable containing at least two different things, such as the sex of a man or a woman ; attitude consisting of agree, neutral or disagree; age which has a value from 0 to 80 for example or money that has a value of 100.000. To distinguish variables with non-variables can be seen from its value. If something is constant then it is not a variable because the constant has a single value.

1.3 Variable Types

In this section the author will explain the terms and examples for independent variables, dependent, moderate, control, and intervening variables.

1.3.1 Independent Variable

The independent variable which is also known as a predictor is a stimulus variable or a variable that affects other variables. The independent variable is a variable whose variability is measured, manipulated, or selected by the researcher to determine their relationship to a phenomenon which is being observed. Another definition saying that an independent variable is a variable thought to affect another variable (Cramer and Howitt, 2006).

In the case of promotion and buying interest variable relationship, promotion is an independent variable that can be manipulated and viewed as an influence on the buying interest variable, for example when we want to study what is the effect of promotion on the buying interest among the consumers at a certain shop. The promotion is called as the independent variable because it affects the buying decision.

1.3.2 Dependent Variable

Dependent variable is a variable that gives reaction / response when connected with the independent variable. The dependent variable is the variable whose variability is observed and measured to determine the effect caused by the independent variable. Another definition says that a dependent variable is a variable that is assumed to depend on, be affected by, or related to one or more independent variables (Cramer and Howitt, 2006).

In the example of the influence of promotion on the buying interest of an XYZ computer, the dependent variable is buying interest. This is because the buying interest of the XYZ computer is affected by the promotion.

The relationship between an independent variable and dependent variable is like the figure 1.1 below

1.3.3 Relationship Between Independent and Dependent Variables

In quantitative research, researchers generally conduct research using more than one variable or at least two variables, including one independent variable and one dependent variable. Both variables are then sought their relationship or influence from one variable to another. To clarify the description, below it will be given an example.

Example 1

- Research hypothesis: There is correlation between leadership style and employee performance
- Independent variable: leadership style
- Dependent variable: employee performance

Example 2

- Research hypothesis: There is correlation between promotion and sales volume
- Independent variable: promotion
- Dependent variable: sales volume

The relationship between the independent and dependent variable can described as follows.

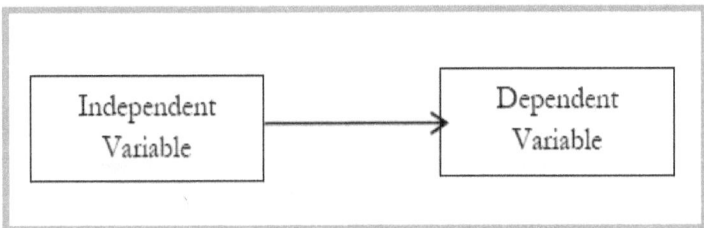

Figure 1.1 Relationship between Independent and Dependent Variables

1.3.4 Moderate Variable

A moderate variable is the second independent variable deliberately chosen by the researcher to determine whether its presence affects the relationship between the first independent variable and the dependent variable. Moderate variables are variables whose variability are measured, manipulated, or selected by the researcher to determine whether they alter the relationship between independent and dependent variable under study.

In case of a relationship between promotion and buying interest, the researcher chooses the moderate variable of price. With the inclusion of price moderate variable, he wants to know whether the magnitude relationship between the two variables change. If it changes then the existence of the moderate variable plays a role, while if not it does not change then the moderate variable does not affect the relationship of both variables under study.

The relationship among the independent, dependent and moderate variables can be seen below.

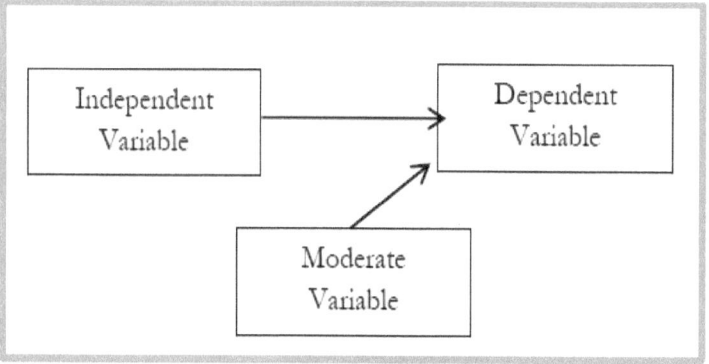

Figure 1.2 Relationship between Independent, Dependent and Moderate Variables

1.3.5 Control Variable

Control variable is a variable that can affect the relationship between the independent and dependent variable in which the effect is controlled or eliminated. The control variable is defined as a variable whose variability is controlled by the researcher to neutralize its influence. The goal is eliminate potential effect from the control variable to the dependent variable.

Example:
- Hypothesis: there is influence of XYZ shoes color to the buying decision among women
- Independent variable: color
- Dependent variable: buying decision
- Control variable: gender

In the case of the above study the control variable is gender. The researcher's assumption is that only women who is affected by the color of shoes.

The relationship among the independent, dependent and control variables can be seen below.

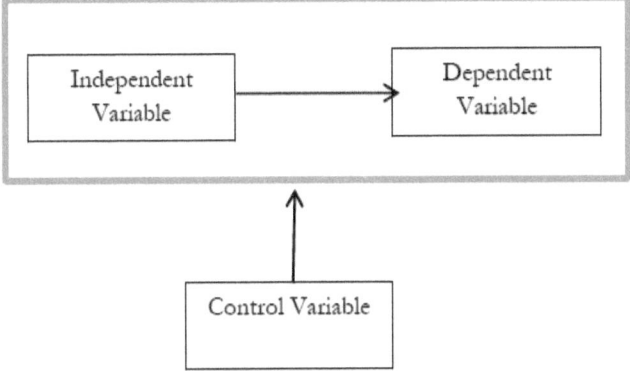

Figure 1.3 Relationship between Independent, Dependent and Control Variables

1.3.6   Intervening Variable

Intervening variable is "a variable thought to explain the relationship between two other variables in the sense that it is caused by one of them and causes the other" (Cramer and Howitt, 2006). The intervening variable is hypothetical meaning that concretely the effect is not visible, but theoretically can affect the relationship between independent and dependent variables under study. Therefore, an intermediate variable is defined as a variable that theoretically influences the relationship of the variable under study but it cannot be seen, measured, and manipulated directly; the effect must be inferred from the influence of independent variable and dependent variable under study.

Example:
- Hypothesis: Good service affects customer satisfaction
- Independent variable: good service
- Dependent variable: customer satisfaction
- Intervening variable: service / product quality

Generally good service will provide high customer satisfaction; although the quality of services will affect the relationship of service and satisfaction variables. Good service does not necessarily give satisfaction to the customer if the quality of service or the product is low.

The relationship between the independent, dependent and intervening variables can seen below.

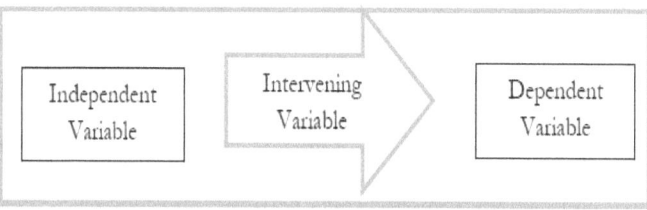

Figure 1.4 Relationship between Independent, Dependent and Intervening Variables

1.4 Variable Relationship Model

Variable relationship model among the five variables. The first model is made by Tuckman (1978:70) as follows.

Data Analysis and Its Interpretation: Application in Marketing Research

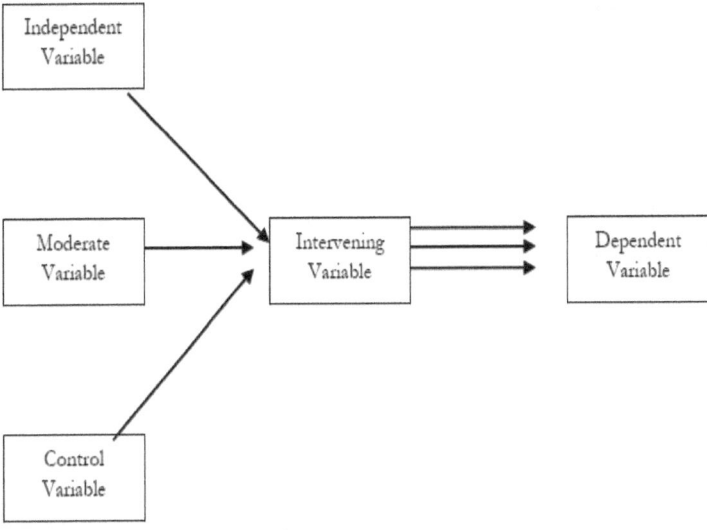

Figure 1.4 Variable Relationship Model by Tuckman

(Source: Tuckman, 1978)

The above model can be read as follows, the main relationship is the independent and dependent variables, the researcher may also consider other variables i.e. moderate and control variables. The relationship of independent and the dependent variable is through an intervening variable. This variable is hypothetical, meaning it is not visible but theoretically exists and affects the relationship between independent and dependent variables.

Another variable relationship model according to Brown (1988:13) are as follows.

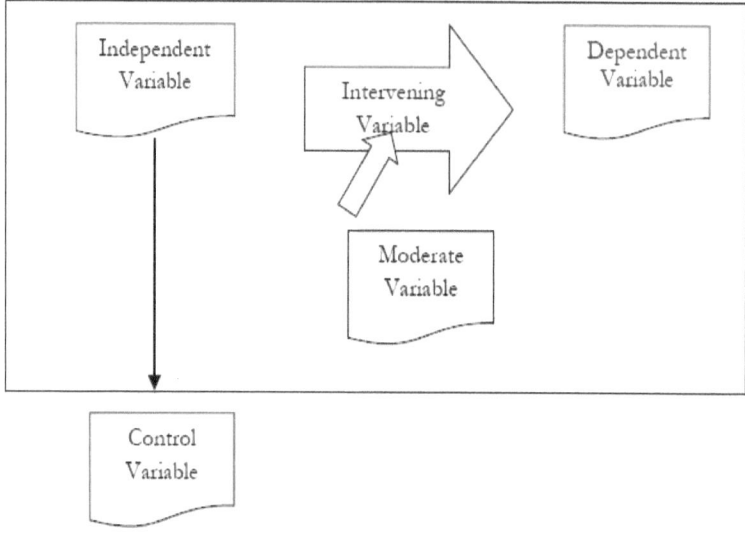

Figure 1.5 Variable Relationship Model by Brown

(Source: Brown, 1980)

The Brown model can be explained as follows: the central relationship in the study is between independent and dependent variables. The arrows further indicate the direction of the researcher's focus and research design, rather than causality. Thus the focus of the variable is dependent variable. In the early stages of the study is conducted only to determine the effect of the independent variable on dependent variable. An intervening variable serves as a label to the relationship of the two variables or the process that links between independent variable and dependent but not as an observed variable meaning that it cannot measured directly. The researcher may also consider another independent variable, i.e. the moderator variable to be used to determine whether there will be a change in the relationship between the independent and dependent variables if the moderator variable is incorporated into the research. The researcher may also control another independent variable if the person wants to neutralize, or eliminate the influence of the control variable.

While according to the author, the relationship model among the principal variables is as follows:

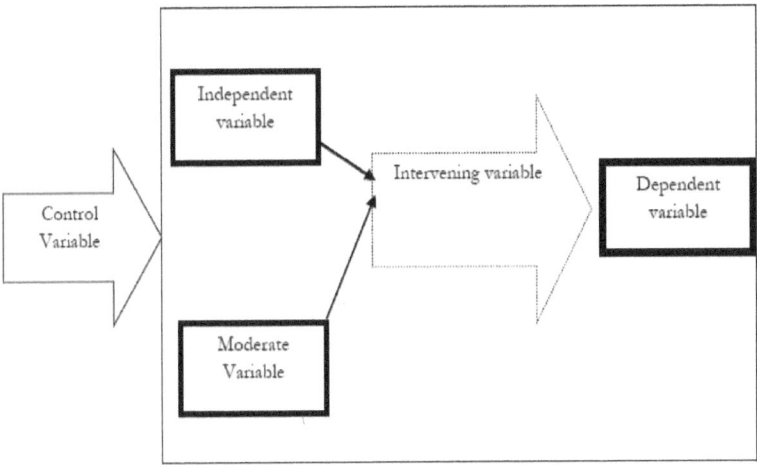

Figure 1.6 Variable Relationship Model

To explain the above picture the author uses the case as follows: the researcher wants to measure the level of customer loyalty of computer X product by using price, computer type, after sales service, and satisfaction level variables.
- Independent variable: price
- Dependent variable: loyalty
- Moderate variable: computer type
- Control variable: gender
- Intervening variable: satisfaction

Description of the above case is as follows: the researcher wants to know whether or not the influence of price on consumer loyalty. Price is independent variable and loyalty is dependent variable. He also considers the existence of other factors that affect the relationship of these two variables, namely the type of computer. The computer type variable is deliberately chosen to determine whether their presence affects the relationship between independent and dependent variables. The researcher intends to neutralize the possibility of the influence of control variable in the form of gender of the customer , therefore it will distinguish between male and female customer satisfaction . The goal is to eliminate the possibility of confusion due to these factors , for example there will be differences in satisfaction among male and female customers in lieu of customer satisfaction alone. In theory, variable satisfaction will affect the relationship between price and loyalty. Then the variable satisfaction serve as an intervening variable.

In the SPSS variable is known as a column. So the number of columns is equal to the number of variables analyzed. While amount of data is referred to as the case entered into the rows

1.5 Measurement

Measurement is putting a numerical values to an observation. For example we give a value of 1, 2 and 3 at the options of disagree, neutral and agree. This is called as a quantitative measurement. This numerical value indicates the amount of something that an observation has. When this applies, we call it quantitative measurement. Besides that there is what is called classification of an observation, for example we classifiy the gender as man and women. This is called a qualitative measurement, naming the defining characteristics of something. Measurement is central in quantitative research because an ovservation should be measured quantitatively.

There are three types of measurement scales IBM SPSS, namely nominal, ordinal, and scale which combine an interval and ratio scale.

1.5.1 Nominal Measurement

The nominal measurement scale is used to classify objects, individuals or groups; for example classifying gender, religion, occupation, and geographic area. In identifying the above things used the numbers as symbols. The classification consists only two possibilities, for example gender is classified into male and female. To measure this variable, the male is quantified as 1 and female 2. We cannot perform arimatic operations with the numbers of 1 and 2, because the numbers only indicate the presence or lack of certain characteristics.

Example:

- Do you agree about the electricity tariff increase?
  Answer: a. yes and b no. Then "yes" is given a value of 1 and "no" is assigned a value of 0

1.5.2 Ordinal Measurement

The ordinal measurement scale provides information about the relative number of different characteristics possessed by a particular object or individual. This level of measurement has nominal-scale information added with a certain relative means of ranking that provides information on whether an object has more or less characteristics but not how many flaws and strengths.

Data Analysis and Its Interpretation: Application in Marketing Research

Example:

- Do you agree about the rise in airfare fares?
  Answer: a. strongly disagree , b disagree, c. hesitate, d. agree, e. totally agree. Then "strongly disagree" option is given a value of 1, "disagree" is 2, "hesitate" 3, "agree" 4 and "agree" 5

1.5.3 Interval Measurement

The interval measurement scale has characteristics such as those of nominal and ordinal scales with other characteristics , that is, in the form of fixed intervals. Thus the researcher can see the magnitude of differences between the characteristics of one individual or object with another. The interval measurement scale is real numbers (numeric). The numbers used can be used to perform arithmetic operations.

Example:

- How many times did you shop at this Supermarket in a month?
  a. 1 time,   b. 2 times,   c. 3 times,   d. 4 times and   e. 5 times

1.5.4 Ratio Measurement

The ratio measurement scale has all the characteristics possessed by the nominal, ordinal and interval scale with the advantages of this scale having an absolute empirical 0 (zero) value. The absolute value of zero occurs in the absence of a characteristic being measured. Measurement ratio is usually in the form of a comparison between one individual or a particular object with another. For example the age range from 7 to 8 years equals between 8 to 9 years range, i.e having a 1 year interval. Age also has absolute value 0 , which is the lowest age is 0 years . Thus the age variable has a scale of measurement ratio because it can be made comparisons as follows: Age 10 years is two times the age of 5 years; the comparison is 2: 1 .

Example:

- What is your weight before and after eating the diet medication? The answer is: Weight before taking 70 kg and weight after taking 60 kg.

In IBM SPSS the interval and ratio measurements are combined into one measurement called scale.

Measurement in IBM SPSS can be summarized as follows:

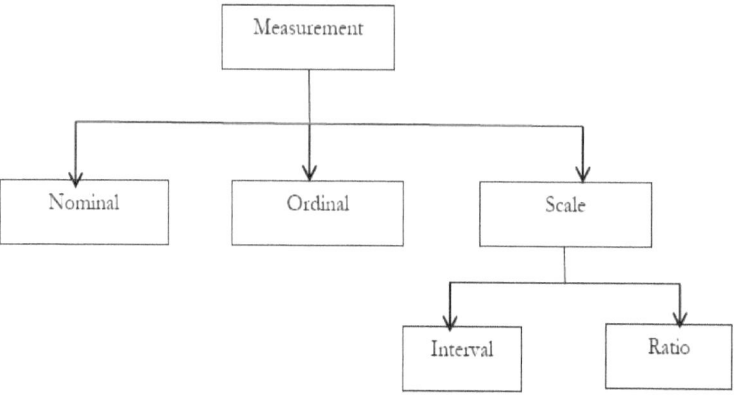

Figure 1.7 Measurement in IBM SPSS

1.6 Confidence Interval

The confidence level or also called a confidence interval or risk level is based on the idea derived from the Central Limit Theorem. The principal idea is derived from the theorem is that if a population is repeatedly drawn by a sample, then the average attribute value obtained from the samples is parallel to the actual population value. Furthermore, the obtained values derived from the drawn samples are normally distributed in the form of true values. The shape of these values will be higher or lower sample values when compared to the population value. In a normal distribution, about 95% of the sample values fall within two standard deviations of the actual population value . In other words, if a 95% confidence level is selected, 95 out of 100 samples will have actual population values within the range of precision as specified earlier. There are times when the sample we get does not represent the actual population value. The confidence level ranges from 99% to the highest and the lowest 90%. In IBM SPSS the confidence level is defaulted by 95%. Common confindence intervals used in the research are 99%, 95% and 90%

1.7 Significance Level / Probability

Significance level or also called probability is the level of precision in terms of sampling error. It is the range at which the exact population value is estimated. This range is often expressed using percentage points, such as 1% or 5%. Therefore, if a researcher finds that 60% of certain company employees used as samples have adopted a recommended working method with a level of accuracy of ± 1%, then the researcher can conclude that between 59% and 61% of the employee of the company becomes the population has already adopted the method. In IBM SPSS the significance is written by default as 0. 05 (5%). Common significance levels used in the research are 1% (0.01), 5% (0.05) and 10% (0.1).

The relationship between confidence interval and significance level can be described as follows:

Confidence Interval (1-α)	Significance Level (α)
99%	1% (0.01)
95%	5% (0.05)
90%	10% (0.1)

Table 1.1 confidence interval and significance level

1.8 Number of Data / Case

In IBM SPSS the amount of data is referred to as the case. How to read it is by looking at the rows. So the number of rows equals the number of cases / data. In IBM SPSS the amount of data is given a symbol of N. SPSS does not distinguish between N (population) and n (sample).

1.9 Hypothesis Testing

This section will provide a basic understanding of the hypothesis, how to create hypotheses and hypothesis testing.

1.9.1 Definition

After the problem is formulated, then the next step is to formulate the hypothesis. What is the hypothesis? There are many definitions of hypotheses that essentially refer to the same meaning. Among them is that the hypothesis is a temporary answer to the problem being studied.

Zikmund (1997: 112) defines the hypothesis as: " Unproven proposition or supposition that tentatively explains certain facts or phenomena; a probable answer to a research question ". According to Zikmund the hypothesis is an unproven proposition or guess which tentatively explains certain facts or phenomena and is also a possible answer to a research question.

While according to Cramer and Howitt (2006:75) hypotheses are allegations or suggestions about the characteristics or traits that may occur which are considered to be a beginning in scientific investigation. Further they say there is no necessity that between the research hypotheses formulated by researchers with statistical hypotheses. In the statistical hypothesis formulation there are always two options that is, the null hypothesis (H0) in which there is no relationship between two variables studied and the alternative hypothesis (H1) which states there is a relationship between two variables studied.

1.9.2 Considerations in Formulating the Hypothesis

In formulating the hypothesis researchers need considerations include:

- It must express the relationship between two or more variables, meaning that in formulating a hypothesis a researcher must at least have two variables to be studied. Both variables should be a predictor and an output. These two variables must have relationship. It can be the predictor correlates with the output or the predictor affects the output.
- It should be stated clearly and directly, meaning that the formulation of the hypothesis must be specific and refers to one meaning and it should not lead to the interpretation of more than one meaning. If the hypothesis is formulated in general, then the hypothesis cannot be tested empirically.
- It should be empirically tested, meaning that it is possible to be disclosed in operational form which can be evaluated on the basis of empirically obtained data from the site. The hypothesis should not reflect the moral elements, values or attitudes.

1.9.3 Types of Hypotheses

According to its form, the hypothesis is divided into three:

- Research hypothesis: a research hypothesis is the basic assumption of a researcher against a problem under study. In this hypothesis the researcher states hypothesis that will then be proven empirically through hypothesis testing by using the data obtained during the research from the site. For example: There is a relationship between the leadership style and the employees' productivity

- Operational hypothesis: an operational hypothesis is an objective hypothesis. This means that the researchers formulate the hypothesis is not solely based on the basic assumption, but also based on objectivity, that the research hypothesis is not necessarily true after being tested using existing data. To that end, the researcher needs a comparative hypothesis that is objective and neutral or technically called the null hypothesis (H0). H0 is used to provide balance to the research hypothesis because the researcher believes in testing later correct or wrong the research hypothesis (H1) depends on the evidence obtained during the research. Example:

H0: There is no relationship between the leadership style and employees' productivity
H1: There is relationship between the leadership style and employees' productivity

- Statistical hypothesis: a statistical hypothesis is a type of hypothesis formulated in the form of statistical notation. This hypothesis is formulated based on researcher's observation of the population in the form of numbers (numeric). For example: H0: $p = 0$; or H1: $p \neq 0$

1.9.4 How to Formulate Hypothesis

How to formulate the hypothesis is with the following stages: 1) formulate a research hypothesis, 2) an operational hypothesis, and 3) a statistical hypothesis.

The research hypothesis is the hypothesis that we make and it expresses in a sentence form and based on the researcher's assumption.

Example 1: Associative hypothesis
Formulation of the problem:

- Is there correlation between leadership style and employee performance?

Research hypothesis:

- There is correlation between leadership style and employee performance

The operational hypothesis is to define hypothesis of the variables in it in order to be operationalized. For example, "leadership style" is operationalized as a way of instructing subordinates. Employee performance is operationalized as the high income of the company. The operational hypothesis is made into two, namely a null hypothesis (H0) which is neutral and an alternative hypothesis (H1) which is based on the researcher's assumption.

Operational hypothesis:
H0: There is no correlation between how to give instructions to subordinates with high - low company revenue

H1: There is correlation between how to give instructions to subordinates with high - low company revenue

Statistical hypothesis:
Statistical hypothesis is the operational hypothesis that is translated into statistical notation according to the measuring instrument chosen by the researcher. In this example assuming a 30% increase in revenue, the hypothesis will be as follows:

H0: $\rho = 0.3$
H1: $\rho \neq 0.3$

Example 2: Descriptive hypothesis
Problem formulation:
- How much is the interest rate increase in X Bank?

Research hypothesis:
- The increase in interest rate at the X Bank is less than the standard

Operational hypothesis:
- H0 = The interest rate increase in the X Bank is the same as the standard
- H1 = The interest rate increase in the X Bank is not the same as the standard

Statistical hypothesis
- H0: $\rho = 5\% (0.05)$
- H1: $\rho \neq 5\% (0.05)$

Its is assumed that the standard increase is equal to 5%.

Example 3: Comparative hypothesis
Problem formulation:
- How is the consumer attitude in Bandung to an increase of train tariff compared with consumer attitude in Yogyakarta

Research hypothesis:
- There is a difference of consumer attitude in Bandung and in Yogyakarta relating to the train tariff increase

Operational hypothesis:
- H0 = There is no difference of consumer attitude in Bandung and in Yogyakarta relating to the train tariff increase

- H1 = There is a difference of consumer attitude in Bandung and in Yogyakarta relating to the train tariff increase

Statistical Hypothesis:
- H0: μ Bandung = μ Yogyakarta
- H1: μ Bandung ≠ μ Yogyakarta

1.9.5 Hypothesis testing

Hypothesis testing is based on two common grounds, namely: the significance level or probability (α) and the confidence level or confidence interval. Based on the significance level in general people use 0.05. The range of significance levels ranging from 0.01 to 0.1. What is meant by significance level is the probability of making a mistake type I i.e. an error to reject the hypothesis when the hypothesis is true. In conducting the hypothesis test there are two hypotheses, namely:
- H0 (null hypotessis) and H1 (alternative hypothesis)

Example:
The average productivity of employees is equal to 10 ($\mu = 10$), then the hypothesis is:
- H0: The employee average productivity is equal to 10
- H1: The employee average productivity is not equal to 10

The statistical hypothesis:
- H0: $\mu = 10$
- H1: $\mu > 10$ For one-tailed test
- H1: $\mu < 10$
- H1: $\mu \neq 10$ For two-tailed test

Some things that must be considered in hypothesis testing is;
- For hypothesis testing we use sample data.
- In testing it will yield two possibilities, i.e. testing statistically significant if we reject H0 and the test is not statistically significant if we accept H0.
- If we use the critical value of t, then if the value of t is getting bigger or away from 0, we will tend to reject H0; otherwise if the t is less or close to 0 we will tend to accept H0.

Using curves to test hypotheses can be described as follows:
   a. A two-tailed test curve. This curve is used for the two tailed test

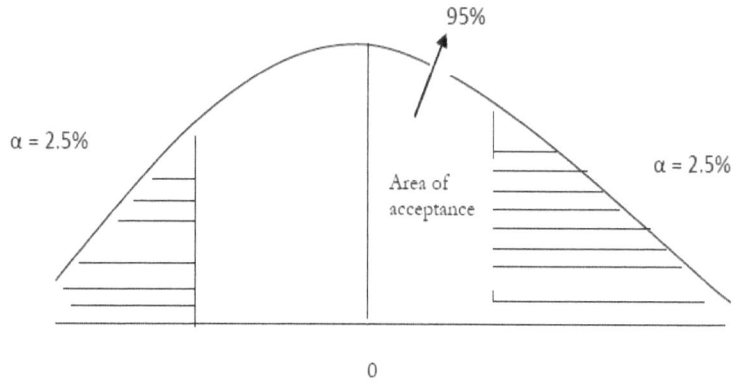

Figure 1.8 A Two Tailed Test Curve

    b. A one-tailed test curve. This curve is used for the one tailed test when the value of observation is positive.

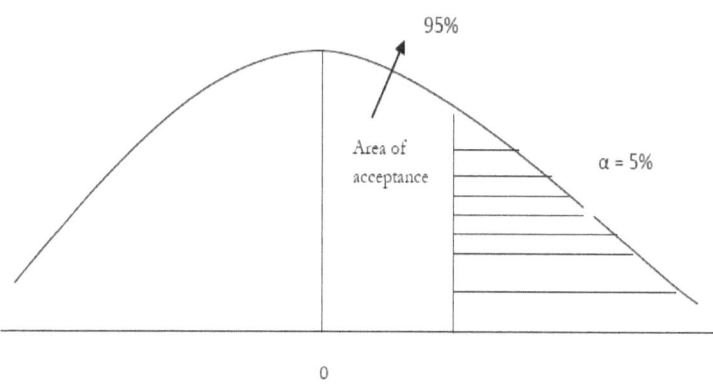

Figure 1.9 A Right Sided One Tailed Test Curve

c. A one-tailed test curve. This curve is used for the one tailed test when the value of observation is negative

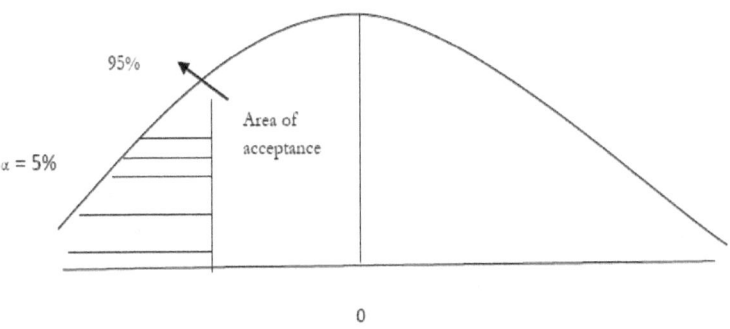

Figure 1.10 A Left Sided One Tailed Test Curve

1.10 Degree of Freedom

The degree of freedom has two different meanings. In this case, in a statistical distribution, this is to give the name of one of its parameters. In relation to the model goodness of fit, the degree of freedom refers to the amount of independent information available to make estimates of other information. Generally we start the number of degrees of freedom with data. The more a procedure or model fits, the smaller the degree of freedom. The calculation of degree of freedom is done through the sample size. The degree of freedom is the measurement of the amount of information from the sample data that has been used. Each statistical calculation is performed from a particular sample, then one degree of freedom is used. Each formula in IBM SPSS has different ways of calculating degree of freedom (DF), for example in Chi Square to calculate DF it uses the formula of (Column-1) x (Row -1); while in the independent sample t test to calculate DF it uses formula of n (number of cases) - 2; for paired sample t test to calculate DF it uses formula of n (number of cases) -1.

1.11 Critical Value

A critical value is used for significance of hypothesis testing. The value at which the statistical tests (the value generating from the observation) must exceed a certain value of the critical value so that the null hypothesis (H0) is rejected. For example the critical value of t with degree of freedom of 12 and the significance level of 0.05 is 1.98. The absolute value of t (the value resulted from the observation) must be greater than 1.98 for H0 to be rejected. The critical value is taken from the critical value table of t while the absolute value comes from the data.

1.12 Statistics Parametric and Nonparametric

Parametric statistics are statistics where the population is assumed to match any measured distribution, generally distributed normally. Statistical parametric inference method is a mathematical procedure for testing statistical hypotheses which assume that the distribution of these variables is being assessed in accordance with the group of normally distributed parameters, for example in IBM SPSS: Linear regression, ANOVA and Pearson Correlation. In contrast, nonparametric statistics relate to nonparametric inference models. This notion also refers to statistics whose interpretations are not dependent on a population matched with any normal distribution, for example in IBM SPSS Spearman Correlation and Chi Square .

1.13 Building a Statistical Model

What is a statistical model? Statistical model is the formalization of relationship between variables in the form of mathematical equations. This statistical model illustrates how one or more random variables are associated with other random variables. What are our uses for building models in statistics? By building a statistical model we can understand how the relation of one random variable to another random variable. For example one variable of promotion has a relationship with another variable called sales. It can be assumed that the variable of promotion correlates with sale.

Building a statistical model also works to capture the real world where the problems we examine originate. Because basically when we build statistical models we actually want to try to describe the real world that technically we call the population into our research. By building the statistical model, we can gain a foothold on where our research is going and how we will design the research.

1.14 Population and Sample

Actually what we examine is an example of the real world. Therefore we must understand what is meant by population. A population is as a representation of the real world and the sample is a part of the population.

Data Analysis and Its Interpretation: Application in Marketing Research

So what is a population? Population is defined as a complete set of analytical units under investigation; while the sample is a subset of the selected set of elements to be studied. In other words the sample is part of the population. As we conduct a research we try to understand the population through the samples we draw from that population. Therefore the sampling method of the population must be done correctly so that our research results can reflect the condition of the population. The relationship between population and sample can be seen in the figure below.

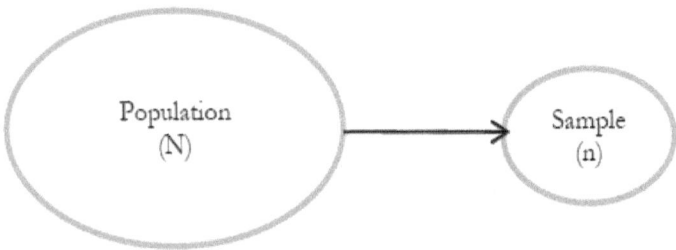

Figure 1.10 Population and Sample

1.15 Data Distribution

One more important factor in research is data; because without the data, a research will not be implemented. How is the data said to be worthy of material analysis in a research? Statistically good data should be normally distributed. What is a normally distributed data? The data is normally distributed in statistics if the data as in the bell shaped form. Ideally the data to be researched should already meet the requirements in advance because if the data does not meet the criteria it then will result in unfavorable research findings. Though statistics also provide formulas for calculating non-compliant data; the task of the researcher is to find the appropriate analytical procedures according to the available data.

1.16 Creating Models That Can Represent the Real World

How is a model good? The model is said to be good if the model can represent the population from which the observed sample originated. With the popular language a model is said to be good if the model can represent the real world. How a model is statistically said to be good depends on every analytical procedure we use. Every procedure in the statistics has a provision or reference to say the model is good or not, such as using the F value or the significance level and the R square value and so forth.

1.17 Counted / Observation / Absolute Value

Counted value is a statistical value, which is also called as an observation or absolute value in a particular procedure, such as the value of t observation in the t test which is the value of the calculation of the data we analyze using the procedure t where the value is a sample value that we will be compared with the critical value. In other words the calculated value is the value of the observation that we will test in a way compared with the critical value derived from the table. For example t value count (t_o) is equal to 2, this is compared with the critical value or t table as much as 1.98.

1.18 Covariate

Covariate is a variable associated with the dependent variable. Because the effect of covariate is often similar to the independent variable; then the covariate is often controlled in a research in order that the bias occurs.

1.19 Metric and Non - Metric Measurement

The measurement scale is required in relation to the data analysis process where the accuracy in determining the measurement scale will affect the accuracy of the results of the analysis. Diversity of data will make it difficult for researchers to determine the suitability between data types and their scale. For that we need a correct understanding of what the measurement scale is. In general the scale of measurement is divided into two, i.e. non-metric and metric measurement scales Non - metric measurement scale is used to measure non-metric data, i.e. data describing differences in types by indicating the presence and void of a characteristic or property. This characteristics are independent and separated by having certain characteristics so that other characteristics are not included in it; for example if a person is a woman then the concerned cannot be a man at the same time. In conclusion these female attributes distinguish themselves with men. Non-metric measurements can be made on a nominal or ordinal scale. Non-metric variables are also called categorical, nominal, binary, or qualitative. In contrast the metric measurement scale is used to measure data that is metric, i.e. data that shows the difference in the number or level of a particular attribute. Therefore, variables that are metricly measured will reflect relative quantities or levels and are suitable for attributes that include the number or size, for example the amount of sales or profits. Metric data measurements are performed using interval and ratio scales. The metric variable is defined as a variable that has a constant unit of measurement; for example if a metric variable is given a scale of 1 - 9, then the difference between 1 and 2 equals the difference between 8 and 9.

Understanding this type of measurement scale will affect two things: first, the researcher must identify the measurement scale for each variable used in his research, so that no errors in measurements are made; secondly, the scale of measurement is important to determine which multivariate techniques will be used for analysis appropriate to existing data, for example metric data should be analyzed by techniques that require metric data.

1.20 Predictor

Predictor is another term of independent variable, that is variable used to predict or explain change value of other variable that respond / which is also called dependent variable.

1.21 Criteria Variable

Criteria variable is also called as a dependent variable, that is a variable that gives response to independent variable which changes our value we measure as effect of the independent variable.

1.22 Factor

It is a variable that is treated as an influencing variable (in some cases in the same with the independent variable) that affects other variables. The difference is that factor is a latent variable that compiles the main / principal variable under study.

1.23 Metric and non-Metric Data

Metric data is also called quantitative data, intervals or ratios. The measurements for this data illustrate not only the ownership of the attributes of an object being studied but also by the number or degree. Non-metric data is also called qualitative data. These data are attributes, characteristics, or categorical properties that describe an object rather than an amount or a level.

1.24 Variate

Linear combinations of multiple variables with empirically determined weights. The variables chosen by the researchers being weighted are determined by multivariate techniques to meet a particular goal.

1.25 Multivariate

The term for designating a statistical procedure that uses more than two variables as an object analyzed

1.26 Variance

Variance is used as one of the instrument descriptor for data distribution and illustrates how far a value lies from the average position. It is also the measurement of the extent to which the values in the population differ from the average value of the population.

1.27 Covariance

Statistical measurements of variants of two random variables that are observed or measured in the same average time period

1.28 Covariance matrix

Covariance matrix is a table to show covariance values

1.29 Categorical

It shows two dichotomous things, for example, a gender variable can be either male or female

1.30 Numeric

Numeric is the same as the actual number, like the integer number

1.31 String

String is the same with a text

1.32 Random Sample

Random sample is taken randomly using probability techniques, a technique that follows statistical procedures.

1.33 Population Vector

Population vector is the desired number of directions in the population

1.34 Normality Assumption

The assumption follows the normal distribution of data where normally distributed data forms a bell-like pattern with an average value being in the middle of the peak.

1.35 Reference Category

Reference category is a non-metric variable category omitted when creating dummy variables (categorical) and it serves as reference points in interpreting the dummy variables.

1.36 Homogenity test

It is testing to find out the similarity of variance and covariance in the compared groups.

1.37 Predicted Value

The predicted value is the value that is calculated from the observed value (data)

1.38 Residual

Residual is a difference between an observation value and predicted value

CHAPTER 2

USING MEANS FOR A COMPARATIVE ANALYSIS

2.1 The First Case : Compare Means

In this procedure we will use data in the average form to analyze cases of hourly wage differences among nurses in different offices and hospitals based on different positions and work experiences. For example in this case we will use SPSS file of hourlywagedata.sav. Using the average score we will see the relationship between wages, experience and position.

To enable the data, we do it in the following way :
- Enable SPSS
- **File> Open > Data**
- Find file location in C> Program Files> IBM> SPSS> Statistics> 25> Samples> English> Select file> Open

To perform an analysis of the data file, the steps are as follows:
- **Analyze> Compare Means> Means**
- Select and move the *Hourly Salary* variable as the dependent variable to the **Dependent Variable** and *Years of Experience* column as an independent variable to the **Independent Variable** column .
- Click **OK**

The result will be as below

Report

Hourly Salary

Years Experience	Mean	N	Std. Deviation
5 or less	18.0416	221	3.86667
6-10	18.9169	460	3.77816
11-15	19.6616	752	3.90528
16-20	20.2876	729	3.82786
21-35	21.2594	539	4.08669
36 or more	21.6342	210	3.61826
Total	20.0159	2911	4.00309

The above results show the salary statistic values based on the level of each experience. Salaries will increase according to the category of experience. Nevertheless, there is the possibility that a large difference in salary is not simply because of their different experiences. There are other factors such as early salary when the employees start working. To add another factor we can add 'layer' in the following way :

- Re-enable the Means dialog box by the way **Analyze> Compare Means> Means**
- Click **Next** to add a **Layer**
- Select and Move the *Nurse Type* variable to the **Independent Variable** column.
- Click **OK**

The result will be as below.

Report

Hourly Salary

Years Experience	Nurse Type	Mean	N	Std. Deviation
5 or less	Hospital	19.0753	147	3.37129
	Office	15.9882	74	3.98762
	Total	18.0416	221	3.86667
6-10	Hospital	19.4846	313	3.35218
	Office	17.7082	147	4.32447
	Total	18.9169	460	3.77816
11-15	Hospital	20.2412	518	3.41065
	Office	18.3784	234	4.57662
	Total	19.6616	752	3.90528
16-20	Hospital	21.1369	471	3.29487
	Office	18.7373	258	4.23293
	Total	20.2876	729	3.82786
21-35	Hospital	21.8601	350	3.48989
	Office	20.1471	189	4.82372
	Total	21.2594	539	4.08669
36 or more	Hospital	22.0641	146	3.14466
	Office	20.6534	64	4.38931
	Total	21.6342	210	3.61826
Total	Hospital	20.6764	1945	3.49582
	Office	18.6859	966	4.58852
	Total	20.0159	2911	4.00309

The above output has the following intent:
In the First Main Line consists of 3 sub-lines, which have the following meanings:
- First row displays statistical values for hospital workers who have experience for ≤ 5 years
- The second line shows the same statistical value for the office nurse with the same level of experience

- The third row of Total which displays the statistical value for the two rows above that are merged.
- Mean Column shows how the average salary differs by based on the level of experience. Although hospitalized people earn more than office nurses at all levels of experience, but the gap narrows from time to time.
- Std Deviation Column shows that the office nurse has a different variation of greater average salary compared to staff nurses.

Conclusion

By adding a classifier variable, then we will be able to quickly observe that the salaries of the nurses consider the experience and the kind nurses who acted as nurses at the hospital or office. This is due to the individual expertise that develops differently in response to the needs of each patient. The office nurse is relatively stable and requires only general skills; while hospital nurses should have special and dynamic skills.

Data Analysis and Its Interpretation: Application in Marketing Research

CHAPTER 3

PRINCIPAL COMPONENT ANALYSIS TO IDENTIFY VALID PREDICTORS

3.1 Definition

Principal Component Analysis (PCA) is a variable-reduction technique that resembles Factor Analysis technique. The main objective of the PCA is 1) to reduce some of the variables which amounts to much into a few variables in a smaller number of so-called principal component that have almost the same variance of the original variables; 2) to detect the relationship between variables with the aim of making the classification of those variables based on the similarity of characteristics in the Factor Analysis using the parameters of the value of MSA (Measure of Sampling Adequacy) on a certain correlation matrix.

In PCA to reduce the number of variables into a group of smaller variables by means of the maximum variation (varimax). This rotation is a way to maximize the value of a variant into a new "created" variable called factor in Factor and Component Analysis in PCA.

3.2 How to Perform PCA Procedures
To perform the analysis with PCA the steps are as follows:

First: Prepare the variables to be scaled down
For example we have 9 independent variables that are assumed to affect the dependent variable. Then we will use PCA analysis to reduce these variables into less amounts of variables. The data is as follows:

sales	product	promotion	distribution	price	quality	process	means	market
6	9	5	6	6	7	8	5	7
5	9	5	4	6	8	8	3	7
3	9	6	5	9	9	9	3	7
8	10	6	5	8	9	9	4	7
9	10	8	6	7	8	9	4	7
5	9	7	6	8	7	10	4	7
4	8	7	5	9	8	8	4	9
3	8	6	5	10	9	8	4	9
2	7	6	6	10	8	8	5	8
5	7	5	4	9	7	7	5	6
4	6	5	5	6	7	7	5	6
9	6	4	6	6	7	7	5	6
10	5	4	6	6	7	8	6	6
10	5	3	5	5	8	8	6	6
8	4	5	4	5	8	6	6	6
7	6	7	7	5	8	6	6	5
6	5	4	7	5	8	6	3	5
6	6	5	7	6	7	5	3	5
9	9	5	8	6	7	5	3	6
9	5	6	9	7	7	5	5	6
8	4	7	7	7	7	6	5	6
8	6	8	6	7	7	6	5	5
7	3	9	6	6	8	4	4	5
7	8	10	5	6	8	3	4	5
6	7	10	5	6	8	6	5	5
6	5	9	6	8	6	6	7	5
5	6	9	6	8	6	6	8	6
5	4	8	6	5	6	5	9	6
8	5	7	6	5	5	5	5	7
6	5	6	6	5	5	5	6	7
7	5	7	5	6	5	8	6	7
7	6	8	8	6	4	8	6	7
8	6	6	7	6	4	7	6	8
5	8	9	9	5	4	7	5	6
6	8	5	1	5	3	7	6	9
7	6	7	2	5	3	7	7	5
8	6	4	3	5	5	7	7	6
9	6	5	3	4	5	6	7	5
6	7	3	4	4	4	6	7	6
6	7	2	4	4	5	6	6	7
8	7	5	4	4	6	5	6	6
9	9	4	5	5	4	5	6	5
9	8	4	6	5	5	7	6	6
4	5	5	5	8	6	7	6	6
5	3	6	6	8	5	8	6	6
6	3	5	5	7	4	8	6	6
7	4	6	6	7	6	8	6	6
5	6	7	8	7	6	7	6	8
4	5	8	5	7	5	7	5	7
6	6	6	6	8	5	7	5	5

Data Analysis and Its Interpretation: Application in Marketing Research

Second: Create a Design Variable in **Variable View** command

Name	Type	Width	Decimal	Label	Value	Missing	Column	Align	Measure	Role
sales	Numeric	8	0	sales	none	none	8	R	Scale	Input
product	Numeric	8	0	product	none	none	8	R	Scale	Input
promotion	Numeric	8	0	promotion	none	none	8	R	Scale	Input
distribution	Numeric	8	0	distribution	none	none	8	R	Scale	Input
price	Numeric	8	0	price	none	none	8	R	Scale	Input
quality	Numeric	8	0	quality	none	none	8	R	Scale	Input
process	Numeric	8	0	process	none	none	8	R	Scale	Input
means	Numeric	8	0	means	none	none	8	R	Scale	Input
market	Numeric	8	0	market	none	none	8	R	Scale	Input

Third: Enter data on SPSS via **Data View** command from data to one to fifty data.

sales	product	promotion	distribution	price	quality	process	means	market
6	9	5	6	6	7	8	5	7
6	6	6	6	8	5	7	5	5

Fourth: To analyze data with PCA procedure

Do the analysis using the following steps:

- Click **Analyze > Dimension Reduction**, then select **Factor**
- Move all variables in the left column to the right variable column
- Select **Descriptives >** check on Initial Solution, Reproduced, Anti-Image and KMO and Bartlett's test of spherecity> Continue
- Select Extraction > Method to select Principal Component Analysis; for Analyze select Correlation Matrix; for Display select Unrotated Factor Solution and Scree Plot; for Extract select Based on Eigenvalue and fill value 1 for Eigenvalues greater than ; ignore other options and click Continue
- Select Rotation> check Varimax> Continue
- Select Scores> check option Save as Variables> Method select Regression > Continue
- Select Options> check option Sorted by size and Suppress small coefficient> Continue
- OK

3.3 Interpretation of Analysis Results
View the results and make interpretations. Only the corresponding output is analyzed while other outputs can be ignored.

Part I: KMO and Bartlett's Test

KMO and Bartlett's Test

Kaiser-Meyer-Olkin Measure of Sampling Adequacy.		,606
Bartlett's Test of Sphericity	Approx. Chi-Square	115,143
	df	36
	Sig.	,000

This section is the output of Measure of Sampling Adequacy which is the value of the first requirement of data feasibility for PCA analysis. KMO Measure of Sampling Adequacy value of 0.607> 0.5 indicates the data feasibility requirement has been met. The minimum value of KMO MSA is 0.5.

Part I I : Measure of Sampling Adequacy (MSA)

Anti-image Matrices

		sales	product	promotion	distribution	price	quality	process	means	market
Anti-image Covariance	sales	,588	-,033	,080	-,147	,216	-,063	-,036	-,061	,142
	product	-,033	,661	-,090	,146	,091	-,022	,238	-,123	-,157
	promotion	,080	-,090	,714	-,189	-,164	-,003	-,098	,177	,088
	distribution	-,147	,146	-,189	,769	-,076	,020	,164	,039	-,012
	price	,216	,091	-,164	-,076	,385	-,141	,047	-,207	-,068
	quality	-,063	-,022	-,003	,020	-,141	,581	,254	,031	,079
	process	-,036	,238	-,098	,164	,047	,254	,507	-,044	-,008
	means	-,061	-,123	,177	,039	-,207	,031	-,044	,577	-,162
	market	,142	-,157	,088	-,012	-,068	,079	-,008	-,162	,627
Anti-image Correlation	sales	,636ª	-,053	,124	-,219	,455	-,108	-,065	-,105	,233
	product	-,053	,548ª	-,131	,205	,180	-,035	,411	-,199	-,245
	promotion	,124	-,131	,473ª	-,256	-,314	-,005	-,162	,275	,131
	distribution	-,219	,205	-,256	,494ª	-,140	,030	,263	,058	-,017
	price	,455	,180	-,314	-,140	,614ª	-,298	,106	-,439	-,138
	quality	-,108	-,035	-,005	,030	-,298	,643ª	,468	,054	,132
	process	-,065	,411	-,162	,263	,106	,468	,594ª	-,081	-,014
	means	-,105	-,199	,275	,058	-,439	,054	-,081	,615ª	-,270
	market	,233	-,245	,131	-,017	-,138	,132	-,014	-,270	,730ª

a. Measures of Sampling Adequacy(MSA)

This section is the MSA output for the variables analyzed, the results are presented starting from the variables with the highest MSA value as follows:
- Market with MSA value of 0.730
- Quality with MSA value of 0.643
- Sales with MSA value equal to 0.636
- Means with MSA value of 0.615
- Price with MSA value of 0.614
- Process with an MSA value of 0.602
- Product with MSA value of 0.548
- Distribution with MSA value of 0.494

When viewed from the requirements then there are only 1 variable having a value smaller than 0.5; namely Distribution variable with MSA value of 0.494.

Data Analysis and Its Interpretation: Application in Marketing Research

Part III : Communalities

Communalities

	Initial	Extraction
sales	1,000	,656
product	1,000	,588
promotion	1,000	,614
distribution	1,000	,490
price	1,000	,800
quality	1,000	,670
process	1,000	,790
means	1,000	,604
market	1,000	,640

Extraction Method: Principal Component Analysis.

Communality is a squared variance that describes how much variance in the measured variables is reproduced by a new variable which is created through a PCA procedure. The value of communality getting closer to 1 is better. The above output is the value of all the variables studied.

Part I V : Total Variance Explained

Total Variance Explained

Component	Initial Eigenvalues			Extraction Sums of Squared Loadings			Rotation Sums of Squared Loadings		
	Total	% of Variance	Cumulative %	Total	% of Variance	Cumulative %	Total	% of Variance	Cumulative %
1	2,731	30,340	30,340	2,731	30,340	30,340	2,264	25,160	25,160
2	1,699	18,883	49,223	1,699	18,883	49,223	1,932	21,468	46,628
3	1,423	15,812	65,035	1,423	15,812	65,035	1,657	18,408	65,035
4	,822	9,134	74,169						
5	,765	8,495	82,664						
6	,576	6,401	89,066						
7	,428	4,758	93,823						
8	,309	3,430	97,254						
9	,247	2,746	100,000						

Extraction Method: Principal Component Analysis.

The above output explains how much variance can be explained in the new variables created by PCA extraction method, that is equal to 65.035%. This value means the creation of new variables can be explained by using the original variables of 65.035%.

Part V : Component Matrix

Component Matrix^a

	Component 1	Component 2	Component 3
price	,804	,141	,367
process	-,622	-,317	,550
sales	-,606		-,530
means	,601	-,490	
market	,600	-,508	,146
quality	,564	,438	-,400
distribution	,147	,683	
promotion	,194	,595	,472
product	,481	-,245	-,545

Extraction Method: Principal Component Analysis.
a. 3 components extracted.

From the extraction results in the form of component matrices, there are three variables are extracted, namely
- Product with a value of 0.481
- Promotion with a value of 0.194
- Distribution with a value of 0.147

CHAPTER 4

ANALYZING RELATIONSHIP BETWEEN MORE THAN ONE INDEPENDENT AND DEPENDENT VARIABLES USING CANONICAL CORRELATION

4.1 Definition of Canonical Correlation

What is a canonical correlation? The canonical correlation according to Hair (2010) can be viewed as a logical extension of multiple regression analysis in order to simultaneously correlate several metric dependent variables with some metric independent variables. The underlying principle is to develop a linear combination of each set of variables (both independent variables and the dependent variables) by maximizing the correlation between the two sets of variables. Therefore in the context of canonical correlations we are allowed to use the term independent variables and dependent variables because the canonical correlation is included in the classification of the dependency test.

Another notion of canonical correlation analysis is a statistical technique used to determine the level of linear association between two sets of variables, where each device consists of several variables. In fact, canonical correlation analysis is an extension of multiple linear regression analysis that focuses on the relationship between two sets of interval-scale variables. The main function of this technique is to see the linearity relationship between the criterion variables (dependent variables) with some independent variables that serve as predictors. For example, a researcher would like to examine the correlation between a set of variables in shopping behavior as a criterion and some variables of personality as predictors. The purpose of this study is to find out how some of these personality characteristics affect shopping behavior, such as making a shopping list, the number of stores visited, and the frequency of shopping in one week.

If illustrated the relationship model between a group of independent and dependent variables will be as below:

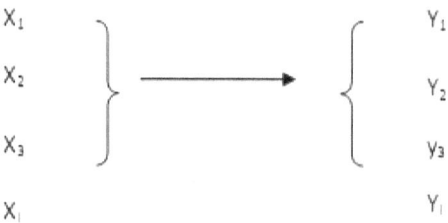

In the group of independent variables are variables of $X_1, X_2, X_3 .. X_i$ and the dependent variables are variables of $Y_1, Y_2, Y_3 .. Y_i$.

4.2 Basic Assumptions

To use this analytical technique the requirements that must be met are:

- The metric independent variable consists of more than two variables, generally the variables have interval scales.

- The metric dependent variable consists of more than two variables, generally the variables have interval scale.

- The relationship between a set of dependent and independent variables is linear. This means that all independent variables influence the direction of all dependent variables, for example in the above example the correlation between the independent variables of the personality used as predictor with the dependent variables used as the criterion is unidirectional. If the value of the personality variable increases, then the value of shopping behavior variables increases as well. If value of personality variable increases while the value of shopping behavior variables become smaller, then this is contrary to the assumption of linearity.

- There should be no multicollinearity in each group of independent variables and dependent variables that will be correlated.

- The sample size can be small and can be large. Small sample sizes can be 100 while large sample sizes can be calculated from the tolerable error rate. For a 5% error rate, then the sample can be as much as 320s being for 1% can be as much as 3200 - an. The disadvantage of using small samples is that sometimes the calculation results are not statistically significant; less represents the relationship between the two variable devices and obscures the relationships between those variables correlated. The advantage of using large samples generally tend to be statistically significant

- The data used is expected to be normally distributed

4.3 Basic Terms

- **Coefficient of canonical correlation:** measures the strength of the overall relationship between two linear canonical variations. One variance for independent variables and one other variables for dependent variables.

- **Canonical cross-loadings:** the correlations of each independent and dependent variable observed with opposite canonical variables

- **Canonical function:** the correlation between two linear canonical variations. Each canonical function has two canonical variables, one for a set of independent variables and the other for a set of dependent variables.

- **Canonical roots:** a squared canonical correlation coefficient that gives estimates of the number of variance divided between each canonical variation of independent and dependent variables which is also called Eigenvalues

- **Canonical variates:** linear combinations representing optimum weighted numbers of two or more variables and formed for independent variables and dependent variables.

- **Canonical loadings:** simple linear correlations between independent variables with and canonical variables respectively.

- **Orthogonal:** a mathematical constraint that specifies that canonical functions are statistically independent or unrelated to each other.

- **Eigenvalues:** same with canonical roots

- **Redundancy index:** the number of variance in the canonical variations described by other canonical variables in canonical functions

- **Variations:** Multiple linear combinations of variables with empirically determined weights. The variables chosen by the researchers being weighted are determined by multivariate techniques to meet a particular goal

4.4 How to Use Canonical Correlation Procedure

For this case we will use SPSS file with bankloan.sav file name contained in program file> folder IBM> SPSS 25> Sample> English> bankloan.sav. Steps using the canonical correlation procedure are as follows:
- Enable SPSS > retreive bankloan.sav data
- **Analyze > correlate > canonical correlation**
- In the column to enter the variable group X Set 1: enter the variables of **Age in years (age)**, **Years with current Employ (employ)**, **Years at current Address (address)**, and **Household Income(income)**
- In the column to enter the variable group Y Set 2: enter the **Debt to income ratio (debtinc)**, **Credit card debt (creddebt)**, and **Other debt (othdebt)**
- Click **OK**

4.5 Interpretation of Analysis Results

The results of the analysis as follows:

Part I: Background of Canonical Correlation

Canonical Correlations Settings

	Values
Set 1 Variables	age employ address income
Set 2 Variables	debtinc creddebt othdebt
Centered Dataset	None
Scoring Syntax	None
Correlations Used for Scoring	3

This canonical correlation involves the association of a set of group variables 1, namely age, employee, address and income with a set of variables of group 2, namely debtinc, creddebt, and othdebt.

Part I: Canonical Correlation

Canonical Correlations

	Correlation	Eigenvalue	Wilks Statistic	F	Num D.F	Denom D.F	Sig.
1	.833	2.275	.303	105.853	12.000	2230.660	.000
2	.074	.006	.994	.854	6.000	1688.000	.528
3	.022	.001	.999				

H0 for Wilks test is that the correlations in the current and following rows are zero

Correlation between sets of group variables 1: age, employ, address and income with a set of 2: debtinc, creddebt, and othdebt variables of 0.833 (this value is taken in Correlation 1 column) and significant because the significance value in the Sig column is 0.000 <0.05. To perform hypothesis testing can be done as follows:

H0: There is no significant correlation between set of group variables 1, namely age, employ, address and income with a set of group variables 2, namely debtinc, creddebt, and othdebt

H1: There is a significant correlation between a set of group variables 1, namely age, employ, address and income with a set of variable group 2, namely debtinc, creddebt, and othdebt

Testing criteria.

If the significance level < 0.05, then there is significant correlation

If significance level > 0.05 then there is no significant correlation

Decision

The significance value in the Sig column is 0.000 < 0.05 thus the correlation between the two groups of variables is significant.

Conclusion

There is a significant correlation between a set of group variables 1, namely age, employ, address and income with a set of variables 2, namely debtinc, creddebt, and othdebt.

Section II: Coefficient of Standard Canonical Correlation

Set 1 Standardized Canonical Correlation Coefficients

Variable	1	2	3
Age	-.016	1.399	.209
Employ	-.114	-.600	-.134
Address	.011	-.760	.956
Income	-.920	-.054	-.323

The correlation between group set 2 with the age variable is -0.016; with employee is equal to -0.114; with address is 0.011 and with income is equal to -0.920.

Set 2 Standardized Canonical

Set 2 Standardized Canonical Correlation Coefficients

Variable	1	2	3
Debtinc	.799	.756	-.602
Creddebt	-.534	-.741	-.988
Othdebt	-.840	.722	.867

The correlation between group set 1 with the debtinc variable is 0.799; with the creddebt variable of -0.534 and with otherdebt of -0.840.

Part III: Coefficient of Non-Raw Canonical Correlation

Set 1 Unstandardized Canonical Correlation Coefficients

Variable	1	2	3
Age	-.002	.174	.026
Employ	-.017	-.089	-.020
Address	.002	-.110	.139
Income	-.024	-.001	-.008

Correlation between set of group 2 with variable age is -0.002; with employee is equal to -0.017; with address is 0.002 and with income is -0.024.

Data Analysis and Its Interpretation: Application in Marketing Research

Set 2 Unstandardized Canonical
Set 2 Unstandardized Canonical Correlation Coefficients

Variable	1	2	3
Debtinc	.119	.112	-.090
Creddebt	-.251	-.349	-.465
Othdebt	-.247	.213	.255

The correlation between group set 1 with the debtinc variable is 0.119; with creddebt variable is equal to -0.251 and with otherdebt is equal to -0.247

Part IV: Canonical Contents
Set 1 Canonical Loadings

Variable	1	2	3
Age	-.511	.585	.555
Employ	-.695	-.120	.110
Address	-.322	-.144	.936
Income	-.996	.003	-.013

The correlation between the Age variable and the canonical variate is -0.511. The correlation between Employ and its canonical variate is -0.695. The correlation between Address and its canonical variate is -0.322. The correlation between the Income with the canonical variate is -0.996

Set 2 Canonical Loadings

Variable	1	2	3
Debtinc	.043	.788	-.614
creddebt	-.664	.114	-.739
Othdebt	-.727	.677	-.115

The correlation between the Debtinc variable with the canonical variate is 0.043. The correlation between the Creddebt variable and the canonical variate is -0.664. Correlation between Otherdebt variables with canonical variate is -0727 .

Section V: Crosslinks

Set 1 Cross Loadings

Variable	1	2	3
Age	-.426	.044	.012
Employ	-.579	-.009	.002
Address	-.268	-.011	.021
Income	-.830	.000	.000

The correlation between the Age variable and the 2 canonical variate of group 2 is -0.426. Correlation between Employ variable and the canonical group 2 is -0.579. The correlation between the Address variable and second group variate is -0.268. The correlation between Income with the 2 group canonical variate is -0.830

Set 2 Cross Loadings

Variable	1	2	3
Debtinc	.036	.059	-.014
Creddebt	-.554	.008	-.017
Othdebt	-.606	.050	-.003

The correlation between Debtinc variable with the first canonical group variate is 0.036. The correlation between the Creddebt variable with the first canonical group variate is - 0, 554. The correlation between Otherdebt variable with the first canonical group variate is -0.606

Section VI: The Proportion of Variance Explained

Total Variance Explained

Component	Initial Eigenvalues			Extraction Sums of Squared Loadings			Rotation Sums of Squared Loadings		
	Total	% of Variance	Cumulative %	Total	% of Variance	Cumulative %	Total	% of Variance	Cumulative %
1	2,731	30,340	30,340	2,731	30,340	30,340	2,264	25,160	25,160
2	1,699	18,883	49,223	1,699	18,883	49,223	1,932	21,468	46,628
3	1,423	15,812	65,035	1,423	15,812	65,035	1,657	18,408	65,035
4	,822	9,134	74,169						
5	,765	8,495	82,664						
6	,576	6,401	89,066						
7	,428	4,758	93,823						
8	,309	3,430	97,254						
9	,247	2,746	100,000						

Extraction Method: Principal Component Analysis.

The proportion of set of variables in group 1 can be explained by a set of variables 2 groups as much as 0.319 or 31.9%. A set of variables in group 2 can be explained by a set of variables in group 1 amounted to 0.225 or 22.5%.

CHAPTER 5

A GOOD MODEL MAKING SIMULATION

5.1 Definition

Simulation is used to create a model that will be used to predict an event with the aim that the resulting output will be at least in accordance with our expectations and objectives to achieve it so we can predict the risks what should be overcome when the program that we make not to be able to run as we expected .

5.2 Case: Using Simulation to Create a Healing Cost Model for Diabet Patients

In this case an insurance company modeled the diabets' healing costs in one year for policyholders so that the company feel confident that the premium they pay will cover all expenses incurred. In its efforts to minimizing the risk that the company tries to exceed the estimated costs that have been paid particularly by the policyholder by asking a distribution fee for the entire population of policyholders who suffer diabetes. One of the questions is what is the threshold for 99% of the population for the cost below that threshold?

To make a statement based on the distribution of costs for the entire population of the policyholder, the researcher must be sure that the concerned has sufficient data to represent the distribution. The data required is the current policyholder's data that has been used to create a predictive model based on the cost per patient. Using the simulation the researcher can simulate as much data as needed and apply the predictive model to the data to obtain the desired cost distribution. The data to be used in this case is the SPSS data file by name diabetes_costs.sav and the model file with the name diabetes_costs.xml, which contains spes ifi k as i from predictable cost model based on the data in diabetes _costs.sav .

5.3 Running the Simulation Procedure in SPSS

To run the simulation the steps are :

1. Open the data file from SPSS with the name diabetes_costs.sav

2. Once the file opens select the menu

Analyze > **Simulation** ... until the simulation dialog box appears as below

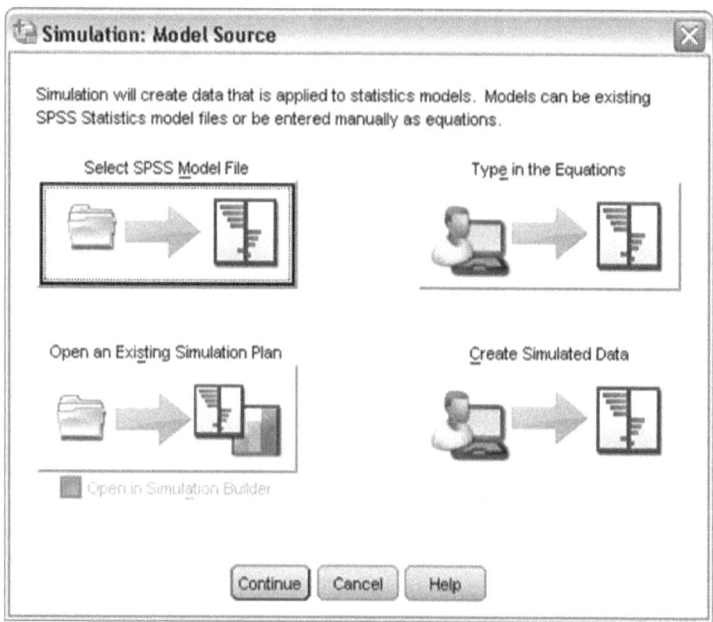

Figure 5.1 Simulation Model Dialog Box

3. Select **SPSS File Model** then click **Continue** .

4. In the dialog box **Select SPSS File Model**, locate directory *Samples* and select the file *diabetes_costs.xml*. A dialog box will appear showing how to create a simulation

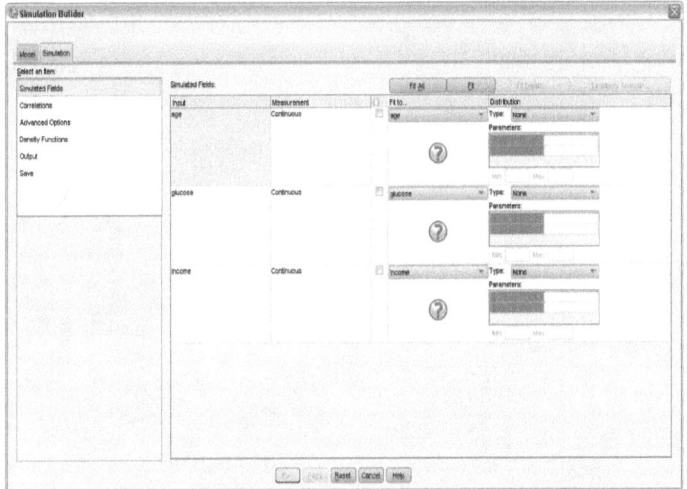

Figure 5.2. Simulation Builder Fields

5. Panel **Simulated Fields** displays a list of all fields which is an input in the prediction model that contains : *age* ; *glucose*, and *income*. To run the simulation we have to create a distribution specification for each input in the model by defining a definite value on the input.

6. Then click selection **Fit All** to automatically match the distribution with existing data. So the results appear as below

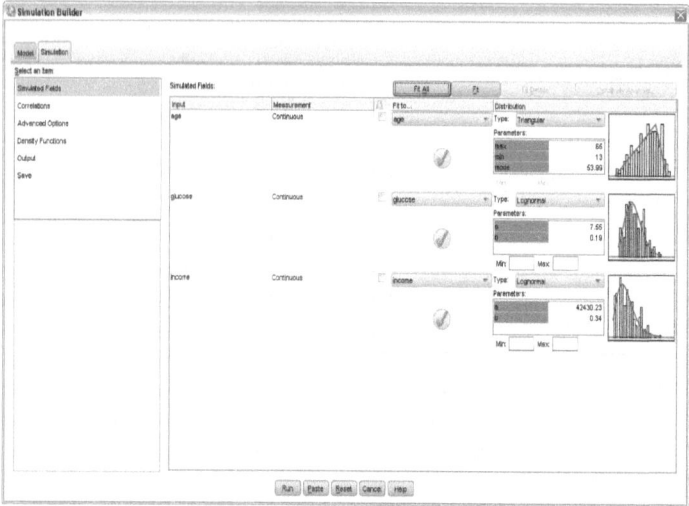

Figure 5.3. Distribution Builder

7. Match results will be shown in columns **Distribution**. The distribution name that best matches the data for each input will be displayed along with all the distribution parameters. For example, datafor *glucose* which best matches the lognormal distribution with the following parameters: a = 7.55 and b = 0.19. The graph adjacent to the distribution specificationi denotes the distribution functioni by displaying in the data histogram.

8. Then click the line for the variable *age* on **Simulated Fields grid** then click option **Fit Details**. The match dialog will appear in detail as below

Data Analysis and Its Interpretation: Application in Marketing Research

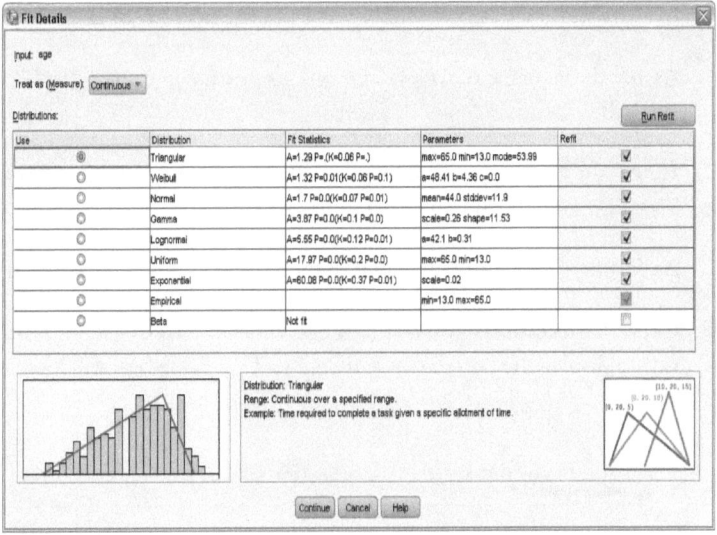

Figure 5. 4 Fit Details Dialog

Using the Anderson-Darling test, the model match results show that the variable 'age' continuity data is the variable that has the distribution that best matches the data. The criterion for using this test is that the smallest statistical value of Anderson-Darling indicates a match with the most appropriate data. This value is coded A in the statistics column and shows a value of 1, 29 in the triangular distribution. The alternative value coded K is Kolmogorov-Smirnov's statistical value. When viewed from the value of significance level / probability (p-value), then the greater the value of significance, that is > 0.05 and approaching1 then the distribution will be closer to the match with existing data. Conversely, if the value of significance is smaller, i.e. < 0.05 and closer to 0, then the distribution is increasingly inconsistent with existing data. From the visual display shows that the triangular distribution is the distribution closest to the data.

9. Click **Cancel** to enter the next variable

When data is simulated for variables '*glucose*' from the appropriate distribution lognormal distribution; then arbitrarily a small positive value will be generated for the lognormal distrubusi rangecovers all the positive values and do not include the value of 0. Glycated hemoglobin levels for people who exposed to diabet commonly found between 5 to 14.

10. Enter the number of 5 on field of **Min** for the variable

of '*glucose*' and insert number of 14 in the field of **Max**.

As for the variable of '*income*', we will use earnings from $ 20,000 up to $ 100.000.

11. Enter the number of 20000 on the field of **Min** for the variable of '*income*' and 100000 on field of **Max**.

As for the variable of '*age*' the minimum value is 13 and maximum 65.

12. Click **Correlations** on option **Select an Item** list in **Simulation tab** to display the correlation simulation. The dialog box will look like this

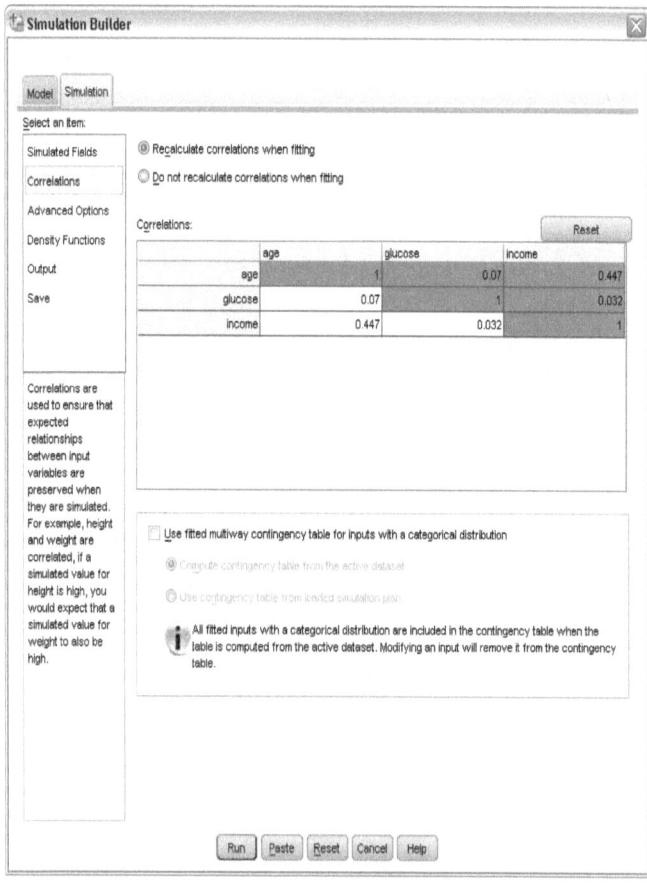

Figure 5.5 Simulation Builder Correlation

Data Analysis and Its Interpretation: Application in Marketing Research

The correlation used in this simulation is Pearson correlation. The correlation result shows as follows:

- The correlation between the '*age*' and '*glucose*' variables is 0.070
- The correlation between the '*age*' and '*income*' variables is 0.447
- The correlation between '*income*' and '*glucose*' variables is 0.032

Correlations will be counted when we click the **Fit All** option or **Fit** on the panel **Simulated Fields**.

13. Click **Output** on the list selection **Select an Item** in the **Simulation tab**. To display the output

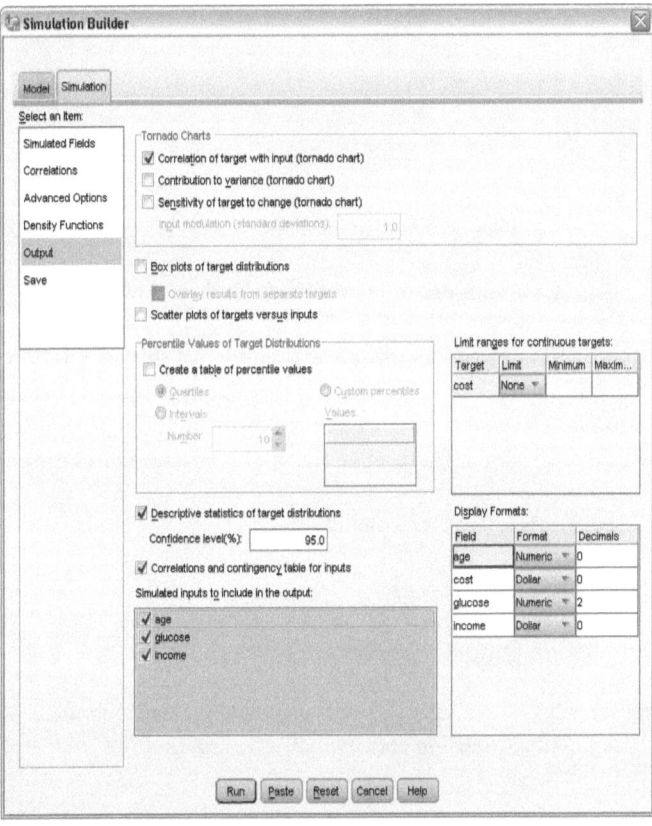

Figure 5.6 Simulation Builder Output

49

To set the output on the **Display Formats** option at the right bottom, change the format for the variable of '*cost*' and '*income*' into the **Dollar** and change number of decimal places to 0. Also give 0 for decimal on the variable of '*age*'. Then select **Save** on the menu list at **Select an Item** on **Simulation tab**.

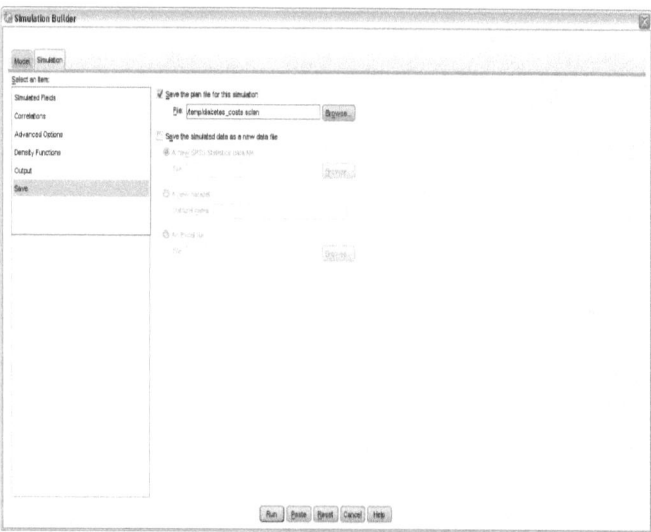

Figure 5.7 Simulation Builder Save

- Check (v) on selection **Save the plan file for this simulation** show that specification i for stimulus of i that will be saved into the simulation plan file. The goal is we can open the file again through **Simulation Builder** or run the file through the option of **Run Simulation**.

- Click **Browse** to locate the file

- Click **Run** to process it

Plan Simulation Output

The output will show simulated details including predictive modeling, distribution used, correlation and anything else required.

Data Analysis and Its Interpretation: Application in Marketing Research

		Simulation		Measurement		Filters	
	Label	Role	Type	Level	Format	Min	Max
age	Age in years	Input	Numeric	Scale	F	.	.
glucose	Glycated hemoglobin level	Input	Numeric	Scale	F,2	>=5.00	<=14.00
income	Household income	Input	Numeric	Scale	DOLLAR	>=$20,000	<=$100,000
cost	Treatment costs	Target	Numeric	Scale	DOLLAR	.	.

Table 5.1 Model Type

The Model Type table displays details about the predictive model on which the simulation is based, as well as any filters specified for the range of values to be simulated for each input

			Parameter Value
age	Triangular	min	13
		max	65
		mode	54
glucose	Lognormal	a	7.547
		b	.192
income	Lognormal	a	42430.232
		b	.340

Table 5.2 Parameter Value Table

Table **Input Distributions** displays distribution associated with each input variable with distribution parameters. For example the distribution of the '*age*' variable is triangular with the lowest parameter value of 13 and the highest 65 with the mode of 54. For the variable of '*glucose*' the lognormal distribution with the parameter of a is 7.547 and b is 0.192 and for the variable of '*income*' the lognormal distribution with a parameter value of a is 42430.232 and b is 0.340.

	age	glucose	income
age	1.000	.070	.447
glucose	.070	1.000	.032
income	.447	.032	1.000

Table 5.3 Correlation Table

Table of **Correlation** shows the correlation between input variables, namely:

- The correlation between the '*age*' and '*glucose*' variables is 0.07 0

- The correlation between the '*age*' and '*income*' variables is 0.447

- The correlation between variable '*income*' and '*glucose*' is 0.032

Maximum cases		100000
Mean within specified precision	Target	cost
	Threshold	1.0%
	Confidence level	95.0%

Table 5.4 Stopping Criteria Table

Table **Stopping Criteria** describes the criteria to be used to determine when discontinuation of simulated generating cases. The criterion is that the simulation will be done until the confidence interval the average target reaches a confidence level of 95% and in 1% of the average value. Maximum cases are 100.000.

Summary

Maximum cases			100000
Total simulated cases			19090
Stopping criteria achieved			Yes
Input filtering	Input: glucose	Minimum value	>= 5.00
		Maximum value	<= 14.00
	Input: income	Minimum value	>= $20,000
		Maximum value	<= $100,000
Cases filtered			3.7%
Simulation Plan File: /temp/diabetes_costs.splan Cases may be filtered because of either targets or inputs that are outside of the specified ranges. Filtered cases are not included in the simulated cases count.			

Table 5.5 Simulation Summary Table

The simulated summary shows the maximum number of cases generated by 100.000 and in this simuliation it generates 19090 cases.

Descriptive statistics

Data Analysis and Its Interpretation: Application in Marketing Research

	Mean	Std. Deviation	Median	Minimum	Maximum	95% Confidence Interval for Mean		Percentiles	
						Lower	Upper	5.0%	95.0%
cost	10538.166	7428.780	8449.245	$1,880	$108,126	10432.785	10643.547	3821.026	23651.829

Table 5.6 Descriptive Statistics of Scale Target Table

This section displays a descriptive statistics summary for the target variable, which is '*cost*'. For example the mean is 10538.166; standard deviation is 7428.780; a minimum charge is 1880 dollars and a maximum charge is 108126 dollars.

	Mean	Std. Deviation	Minimum	Maximum
age	44.219	11.041	14	65
glucose	7.737	1.451	5.00	13.97
income	44814.264	14645.524	$20,002	$99,924

Table 5.7 Descriptive Statistics of Scale Input Table

This section shows descriptive statistics of input variables, namely 'age', 'glucose', and 'income'. For example for the 'age' variable. The age average of the policyholder is 44.219; the lowest age of 14 years and the highest age is 65.

Correlation Tornado Graph

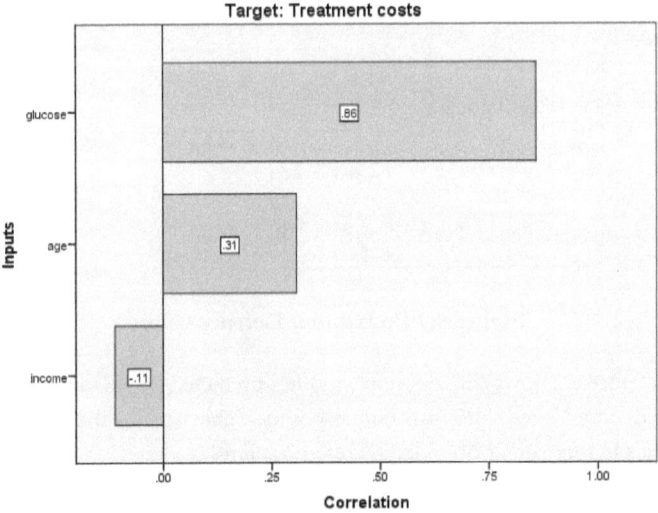

Figure 5.8 Describes the correlation between target and input variables

The graph above shows the Pearson correlation between the target variable of 'cost' and the input variables i.e. 'age', 'glucose', and 'income'. If the value of correlation coefficient is:

- The correlation between variable of 'cost' and 'glucose' is 0.86

- The correlation between the variable of 'cost' and 'age' is 0.31

- The correlation between variable of 'cost' and 'income' is 0.11

The conclusion is the cost of diabetic healing correlates highest with glycated hemoglobin levels in individual insurance policyholders .

Graph of Probability Density

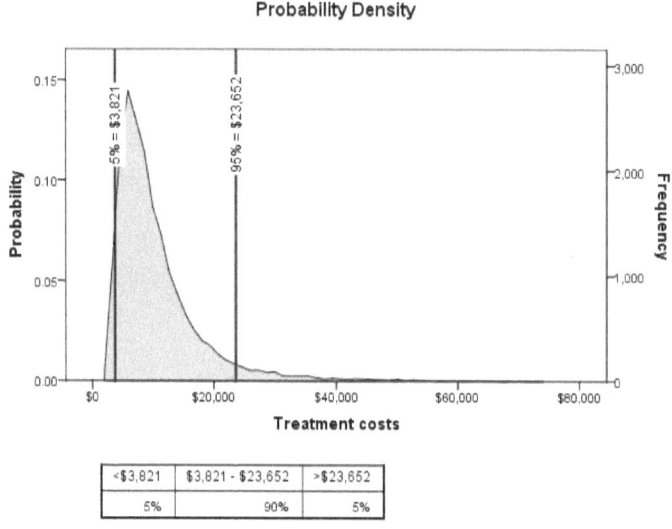

Figure 5.9 Probability Density Chart

The table above shows distribution models predicted target variable. Reference lines placed at 5 % and 95% distribution points. Table below the graph presents probabilities in three areas bounded by reference lines.

Part 2. Probability Density in Graphboard Editor

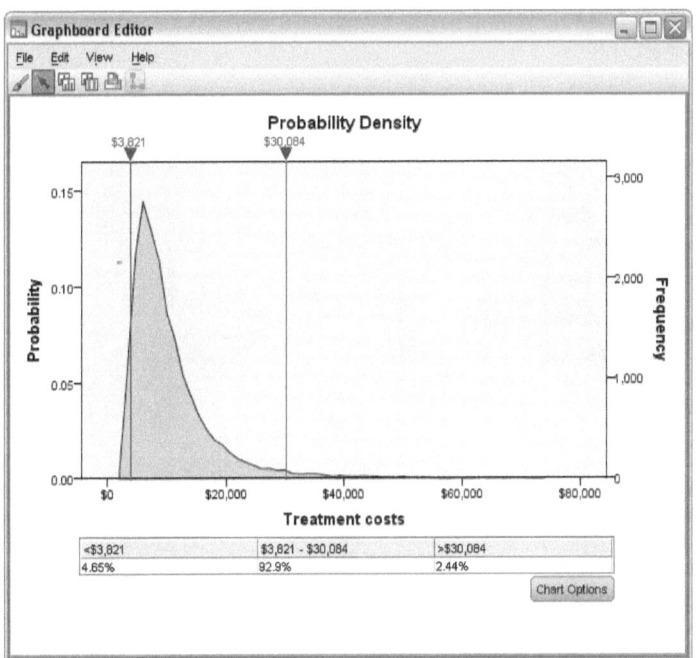

Figure 5.10 a picture of probability density in the form of a graphboard editor

We can make changes to the above graph. For example we can determine probabilitys which relates to healing costs that exceed 30000 dollars by drawing a line next near to $ 30,000. We can arrange the position by using the line or by using the real value of the dialog box of **Chart Options.** We can answer the question of the cost threshold for 99% of the population below that value by simply drawing the right-hand line until the probability that the third column in the table close to 1%. With the right-hand line in this position, then the label above the line will indicate the cost that is sought.

If we want to use the actual value listed on Chart Options, then the way is :

- Click **Chart Options**. It will appear the dialog box as below.

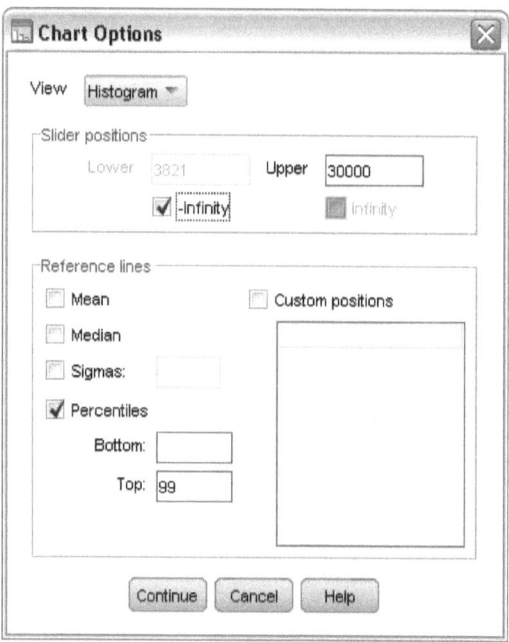

Figure 5.11 Chart Options Dialog

- Choose **Percentiles** and enter a value 99 on **Top field**. To be more precise about the probability of healing costs that surpass the $ 30,000 input value of 30000 in the **Upper field,** on **Slider position** option .

- Select **Infinity** on **Slider position**

- Choose **Histogram** from the menu **View** .

- Click **Continue**

The result will be as below.

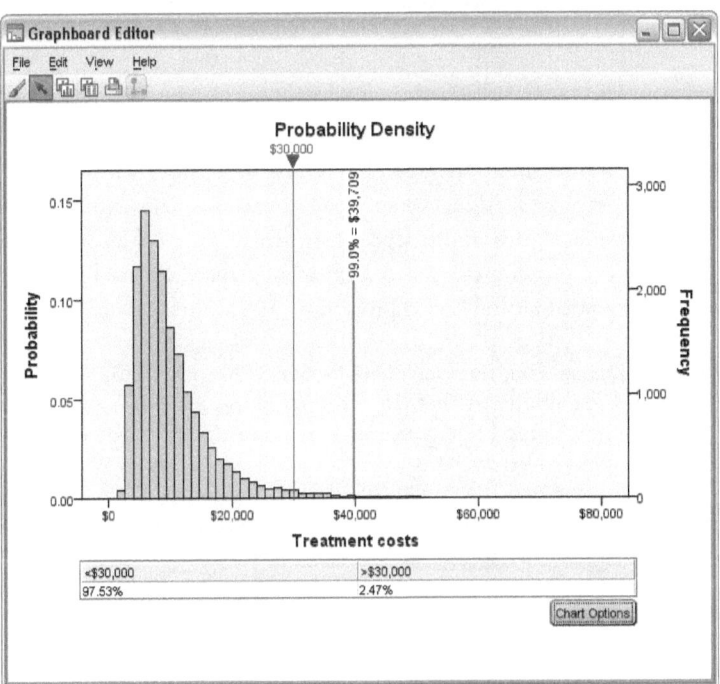

Figure 5.12 Probability Density

The results show a reference line pointing to 99% of the population has a healing cost of < $ 39,709. If the reference line is shifted towards 30000, then we can say that by 2.47 % of the population has a healing cost of > $ 30,000.

CHAPTER 6

HOW TO ANALYZE THE DATA HAVING DIFFERENT MEASUREMENT SCALES

6.1 Introduction

To analyze data having different measurement scales, we can use the General Linear Procedure (GLM) in IBM SPSS. The GLM procedure is a procedure based on the General Linear Model where factors and covariates are assumed to have a linear relationship with the dependent variable. In a GLM procedure there are two types of procedures, namely Univariate General Linear Procedure (GLM) and Multivariate General Linear Procedure (MGLM).

The first part of this session, the Univariate General Linear Procedure (GLM) procedure that will be discussed first. In order to know the GLM procedure well, we need to know some basic concepts used in GLM. Some of the basic concepts will be explained below.

Factor: In univariate GLM a factor is a predictor (another word of an independent variable) in the model in the form of categorical variable. Each level in a factor can have a different linear effect on the value of the dependent variable. There are two types of factors, namely fixed and random factors in GLM.
- **Fixed-effects factors** are variables in which the importance values are all represented in the data file.
- **Random-effects factors** are variables where values in the data file can be considered as random samples of a larger population of values. Random factors are useful to explain the variability that exists in a dependent variable.

For example a wholesaler is interested to see the effects of 5 different coupon types on the amount of customer spending. In some store locations, the coupons are given to customers who frequently visit the site; while 1 coupon is chosen randomly to be given to each customer. This type of coupon is a fixed effect because the company is interested in the particular coupons. Average location of the store is a random effect because of the location that is used is a sample of the population of greater interest. In addition there are variations from one store to another in relation to the amount of expenditure spent by the customer; where company are not directly interested in such variations in the context of this issue.

Covariate: A covariate is an independent variable that can have an effect on the dependent variable beside the main independent variable in the model under study. In GLM the covariate is treated as a predictor (or independent variable) which is a an independent variable that has scale value. In combination, the factor and the covariate values are assumed to be linearly correlated with the dependent variable values.

Interaction: By default the univariate GLM procedure yields a model with factorial interaction, which means that each combination of factor levels can have different linear effects on the dependent variable. We can also add specifications to the interaction between the factor and the covariate if it is assumed that the linear relationship between the covariate and the dependent variable changes at different levels on a single phase .

6.2 Assumptions

For the purpose of hypothesis testing related to parameter estimation, GLM has the following assumptions:
- Errors are independent of each other within all variables in the model. This assumption should not be violated to produce a correct result.
- The error variability is constant in all cells (tiers). This is especially important when there are unequal cell sizes, where there are different numbers of observations across all factor-level combinations .
- The error has a normal distribution with an average of 0.

6.3 Basic Terms

Some of the basic terms that are often used in this procedure are:
- **GLM:** It is abbreviation of General Linear Model
- **MGLM:** It is abbreviation of Multivariate General Linear Model
- **Independent variable:** a variable that is identified as the variable that affects other variables, or causes of fluctuations in the value of the variable that responds to it. Sometimes it is called as a predictor.
- **Predictor :** a predictor is equal to an independent variable, that is a variable used to predict or explain value change of other variables that respond. It is also called as a dependent variable.
- **Dependent Variable:** a dependent variable is a variable that responds to the independent variable whose value changes we measure as a result of the influence of the independent variable. It is also called as a response variable.
- **Factor:** a factor is an independent variable which is either fixed factor and random factor, which is a variable that is treated as a variable that affects other variables. The difference is that a factor is latent variable that compiles the main / principal variables under investigation.

- **Variate**: Linear combinations of multiple variables with empirically determined weights. The variables chosen by the researchers being weighted are determined by multivariate techniques to meet a particular goal.
- **Multivariate**: a term for designating statistical procedures that use more than two variables as the object being analyzed
- **Variance**: it is used as one of the instrument descriptor for data distribution and describes how far a value lies from the average position. It is also the measurement of the extent to which the values in the population differ from the average value of the population.
- **Covariance**: measurement of variance statistics of two random variables observed or measured in the same average time period
- **Covariance matrix**: a table to show covariance values
- **Quantitative data** : data in the form of actual numbers (numeric) can be either continuous or discrete numbers.
- **Metric scale (scale values):** metric measurement scales are used to measure data that is metric, i.e. data that show differences in the amount or degree in a particular attribute. Therefore, the variables measured in terms of metrics will reflect relative quantities or levels and are suitable for attributes that include the number or size, for example the number of sales or profits. Metric data measurements are performed using interval and ratio scales. The metric variable is defined as a variable that has a constant unit of measurement; for example if a metric variable is given a scale of 1 - 9, then the difference between 1 and 2 equals the difference between 8 and 9.
- **Non-metric scale:** a non-metric measurement scale is used to measure non-metric data, i.e. data describing differences in type by showing the presence and vacancy of a characteristic or property. This characteristic is independent and separated by having certain characteristics so that other characteristics are not included in it; for example if a person is a woman then the concerned cannot be a man at the same time. In conclusion these female attributes distinguish themselves with men. Non-metric measurements can be made on a nominal or ordinal scale. Non-metric variables are also called categorical, nominal, binary, or qualitative variables.
- **Categorical** : it shows two dichotomous things, for example, a gender variable can be either male or female
- **Numerical**: the same as the actual number, like the integer
- **Random samples**: samples taken randomly using probability techniques, a technique that follows statistical procedures of a particular population .
- **Population vector**: the desired number of directions in the population
- **Normality assumption**: an assumption that follows the normal distribution of data where normally distributed data forms a bell-like pattern with an average value of 0 and is at the center of the peak of the bell curve.

Data Analysis and Its Interpretation: Application in Marketing Research

- **Reference Category**: a category of non-metric variables omitted when creating dummy variables (categorical) and serves as a reference point in interpreting the dummy variable.
- **Homogenity test:** testing to find out the similarity of variance and covariance in the comparison group
- **Predicted Value** : a value that is predicted from the observation value (data)
- **Residual**: a difference between the predicted value and the observed value
- **Eta**: a value similar to the Pearson correlation coefficient used to describe the strength of the relationship between two variables. The difference is if the Pearson correlation coefficient reflects the relationship of two variables which are independent each other while eta reflects the relationship of two variables which are dependent, i.e. there should be variables identified as independent and dependent variables. Thus the eta squared value is the value that has similarity with r square value (r^2) which is called as coefficient of determination which functions is to explain how much variation of dependent variable value that can be explained by using the independent variable.

6.4 Cases

In this example we will use the IBM SPSS file. A grocery store conducts a customer-related survey of purchasing habits. From the results of the research and the amount of each customer's expenditure in the previous month, the store wants to know if the frequency of customers doing the shopping related to the amount of money spent in each month is controlled by the customer's gender. This information is collected in file by name of grocery_1month.sav .

6.5 Using univariate GLM to perform a two factor variance analysis

To perform the analysis with GLM the steps are:
- **Analyze > General Linear Model > Univariate ...**
- choose *Amount spent* as the dependent variable .
- choose *Gender* and *Shopping style* as fixed factors.
- Click **Plots.**
- choose *style* as horizontal axis variable.
- choose *gender* as separate lines variable .
- Click **Add** .
- Click **Continue** .
- Click **Post Hoc** on GLM Univariate dialog box.
- choose *style* as a variable where **post hoc tests** generated .
- choose **Tukey** on **the Equal Variances Assumed group** .
- Click **Continue.**
- Click **Options** on GLM Univariate dialog box .
- choose *gender * style* as basis to display the mean.

- choose **Descriptive statistics**, **Homogeneity tests**, **Estimates of effect size**, and **Spread vs. level plot** on the Display group.
- Click **Continue**.
- Click **OK** on GLM Univariate dialog box

6.6 Analysis and Interpretation

Part I: descriptive statistics which are a combination of factors in the model.

Descriptive Statistics

Dependent Variable: Amount spent

Gender	Shopping style	Mean	Std. Deviation	N
Male	Biweekly; in bulk	413.0657	90.86574	35
	Weekly; similar items	440.9647	98.23860	120
	Often; what's on sale	407.7747	69.33334	30
	Total	430.3043	93.47877	185
Female	Biweekly; in bulk	343.9763	100.47207	35
	Weekly; similar items	361.7205	90.46076	102
	Often; what's on sale	405.7269	80.57058	29
	Total	365.6671	92.64058	166
Total	Biweekly; in bulk	378.5210	101.25839	70
	Weekly; similar items	404.5552	102.48440	222
	Often; what's on sale	406.7681	74.42114	59
	Total	399.7352	98.40821	351

If seen in the column of Total then look at the Shopping Style variable in average customer spend $ 378.52 on the category every two weeks, while the expenditure category weekly is $ 404.55; while in the category of "frequent" customers spend as much as $ 406.76 based on the variable of *Gender*. Male customers spend $ 430.30 compared to female customers which spent $ 365.66.

Section II : Equality of Error Variance

Variance equality test is done by using significance value at Levene's Test at following output.

Levene's Test of Equality of Error Variances[a]

Dependent Variable: Amount spent

F	df1	df2	Sig.
1.157	5	345	.330

Tests the null hypothesis that the error variance of the dependent variable is equal across groups.

a. Design: Intercept + gender + style + gender * style

To test the equality of the error variance is done through the following stages:
Hypothesis
H0: Error variance of the dependent variable is the same

Data Analysis and Its Interpretation: Application in Marketing Research

H1: Error variance of the dependent variable is not the same

Criteria
If the significance level < 0.05; then H0 is rejected and H1 is accepted
If significance level > 0.05, then H0 is accepted and H1 is rejected

As seen at the output of significance level value on the Sig column shows the value of 0.330 > 0.05; then H0 is accepted. The conclusion is the error variance of the dependent variable is the same. Thus there is no violation of the assumption in this procedure.

Part III : Spread-versus-level output plot

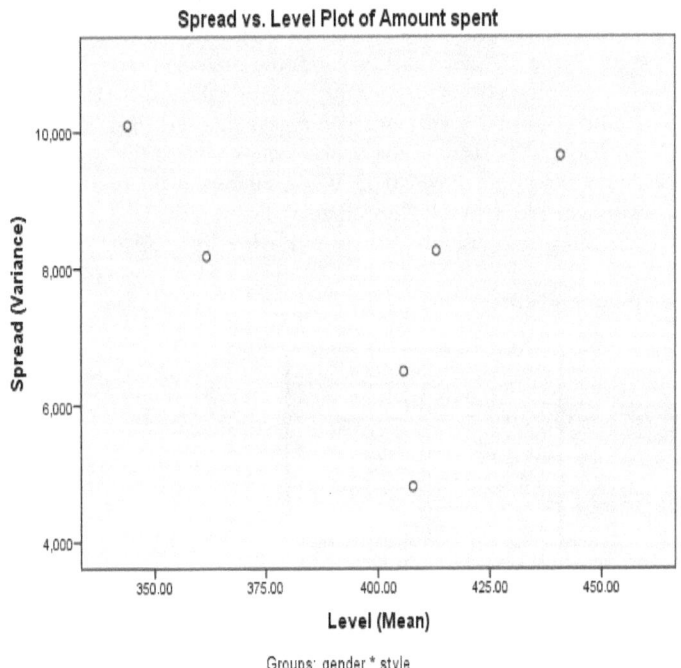

The distribution of the data in the graphic output above does not indicate the existence of certain patterns of relationship between the average value and the standard deviation and thus no violation of assumptions. This output reinforces the equality test of error variance on Levene testing above.

Part IV : Post Hoc Testing

Multiple Comparisons

Dependent Variable: Amount spent
Tukey HSD

(I) Shopping style	(J) Shopping style	Mean Difference (I-J)	Std. Error	Sig.	95% Confidence Interval	
					Lower Bound	Upper Bound
Biweekly; in bulk	Weekly; similar items	-26.0342	12.61108	.099	-55.7191	3.6507
	Often; what's on sale	-28.2471	16.25946	.193	-66.5198	10.0256
Weekly; similar items	Biweekly; in bulk	26.0342	12.61108	.099	-3.6507	55.7191
	Often; what's on sale	-2.2130	13.47525	.985	-33.9320	29.5061
Often; what's on sale	Biweekly; in bulk	28.2471	16.25946	.193	-10.0256	66.5198
	Weekly; similar items	2.2130	13.47525	.985	-29.5061	33.9320

Based on observed means.
The error term is Mean Square(Error) = 8463.939.

This section helps us determine the significance of a factor. The point is whether the factor has a significant impact or not on the dependent variable under study. Post hoc testing shows the mean difference in the predicted model for each pair at the factor level. The first column shows different post hoc tests. The second column shows the pairs of factors that are being tested. There is a significant difference due to the effect of factors on the dependent variable if the significance value in the Sig column shows a value < 0.05. When viewed in the above output there are no values which are < 0.05. Thus there is no significant difference in shopping habits in the category of weekly, weekly and "often" shopping.

Section V: Homogeneous Subset
Tukey HSD[a,b,c]

Shopping style	N	Subset
		1
Biweekly; in bulk	70	378.5210
Weekly; similar items	222	404.5552
Often; what's on sale	59	406.7681
Sig.		.116

Means for groups in homogeneous subsets are displayed.
Based on observed means.
The error term is Mean Square(Error) = 8463.939.
a. Uses Harmonic Mean Sample Size = 83.941.
b. The group sizes are unequal. The harmonic mean of the group sizes is used. Type I error levels are not guaranteed.
c. Alpha = .05.

In homogenous subset testing can be done by looking at the values grouped in the column. It can be seen above that all the mean values that indicate spending in shopping in the two-week, weekly and "frequent" shopping categories are placed on the same one column, that is on Subset 1. This means that all levels of factors do not differ significantly respectively. The conclusion is that attempts to encourage customers to shop more often than their habits do not result in

Data Analysis and Its Interpretation: Application in Marketing Research

significant results because in reality there is no difference in spending at all three levels of the factors under study.

Section VI: Estimated Marginal Rate

Gender * Shopping style
Dependent Variable: Amount spent

Gender	Shopping style	Mean	Std. Error	95% Confidence Interval	
				Lower Bound	Upper Bound
Male	Biweekly; in bulk	413.066	15.551	382.479	443.652
	Weekly; similar items	440.965	8.398	424.446	457.483
	Often; what's on sale	407.775	16.797	374.738	440.812
Female	Biweekly; in bulk	343.976	15.551	313.390	374.563
	Weekly; similar items	361.720	9.109	343.804	379.637
	Often; what's on sale	405.727	17.084	372.125	439.329

The above output shows the estimated marginal mean value and standard error of the dependent variable of *Amount spent* on a combination of factor of *Gender* and *Shopping style*. This table is useful for exploring the possible interaction effects of both factors. In this case example, male customers who make purchases every week spend approximately as much as $ 440 , 96; while people who make purchase more often are expected to spend as much as $ 407.77. Meanwhile, female customers who spend every week are expected to spend as much as $361.72; while people shopping are often expected to spend as much as $ 405.72. Thus there is a difference between customers who do weekly and more frequent purchases depending on the gender of the customer . This fact proves that the interaction effect between factors of *Gender* and *Shopping style* does exist. If there is no interaction, then we expect that the difference between shopping styles remains constant between male and female customers. This interaction will become clearer as we look at the customer profile graph below.

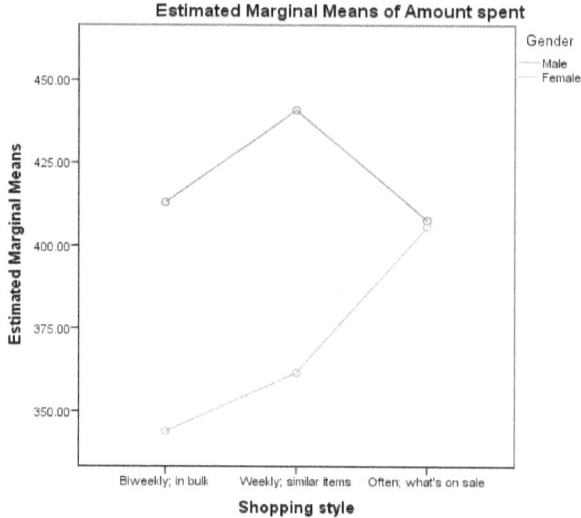

This profile graph is representative of the visual table of the marginal mean above. The factor of *Shopping style* is on horizontal axis line. While a separate line produced is the level of each factor of *Gender*. If there is no interaction effect then the lines in the table above will be parallel. In the graph above, we see that the difference between customers who shop in the weekly category is often more done by female customers. This is seen on the line indicating a decreasing male customer being for an ascending female customer. The interaction effect is not due to chance but because in the test between the effects of the subjects showed a significant statistical value.

Data Analysis and Its Interpretation: Application in Marketing Research

Section VII: Testing Inter-Subject Securities

Tests of Between-Subjects Effects

Dependent Variable: Amount spent

Source	Type III Sum of Squares	Df	Mean Square	F	Sig.	Partial Eta Squared
Corrected Model	469402.996a	5	93880.599	11.092	.000	.138
Intercept	39359636.387	1	39359636.387	4650.274	.000	.931
Gender	158037.442	1	158037.442	18.672	.000	.051
Style	33506.210	2	16753.105	1.979	.140	.011
gender * style	69858.325	2	34929.163	4.127	.017	.023
Error	2920058.824	345	8463.939			
Total	59475118.440	351				
Corrected Total	3389461.820	350				

a. R Squared = .138 (Adjusted R Squared = .126)

The above output is an analysis of the variance. Each causal factor in the model, with an overall plus models, tested for its ability to account for miscellany which is contained in the dependent variable. The significance value for all sections in each "Source" column shows the significance value in the Sig column smaller than 0.05; except a variable of *Style*. Thus we can conclude that any causal factors start from the *Corrected Model* which shows the feasibility of the model where the model of relationship between the variables that we make is feasible because the significance value in this section of 0.000 <0.05; *Intercept* which represents a constant value when the predictor variable of 0 is also significant because its significance value is 0.00 <0.05; a *Gender* variable as a factor also has a significance value of 0.00 < 0.05 and interaction between variables *Gender* and *Style* also has a significance value of 0.017 < 0.05 indicates statistically significant. Of all the parts except the *Style* variable with a significance value of 0.140 which has no significant effect to the dependent variable of shopping expenditure (*Amount Spent*).

At a section of *Partial Eta Squared* shows the value of practical significance each section, based on *sum of squares* that can be explained by a factor (independent variable), to the amount of variation caused by factors that causes and the remaining variability of the error or variations caused by factors other than *Gender* and *Style*. The greater the value of *partial eta squared*, i.e. close to 1 indicates the number of variations that can be explained in the model by the causal factor (independent variable). Yet every factor causing is statistically significant but does not have a major impact on the amount of money spent (*Amount spent*).

Conclusion

In the above case, post hoc testing does not reveal differences in spending expenditures among customers who shop twice every week and who shop more often. Even so, the mean value of the estimated marginal and graphs can reveal their profile interaction between the two factors, implying that the male customers who only shop once a week is more profitable customers than all the customers who often shop; otherwise this pattern becomes reversed for female customers. The significance of this interaction effect is confirmed from the ANOVA table results.

6.7 Using Univariate GLM to Conduct Covariance Analysis

The next case relates to government efforts that run the program to see whether the program will help the community in obtaining a better job. The program is controlled by the salary factor. Most of them who are included as the program participants are randomly selected to participate sign up for the program; while others do not. For this case all data is in the file *workprog.sav*.

To perform the GLM analysis procedure the steps are as follows :

- **Analyze** > **General Linear Model** > **Univariate**
- If previous data is still there click **Reset** first.
- Select a variable of *Income after the program* as the dependent variable.
- Select a variable of *Status program* as the fixed factor.
- Select a variable of *Income before the program* as a covariates.
- Click **Model**
- choose **Custom** a model type.
- Block simultaneously *prog* and *incbef* in the list **Factors and Covariates** and move it to the **Model** column.
- choose **Interaction** from list **Build Term (s)** for the model.
- Click **Continue**
- Click **Options** on the box GLM dialog.
- choose **Estimates of effect size** on **the Display group**.
- Click **Continue**
- Click **OK** in the Univariate GLM box

Data Analysis and Its Interpretation: Application in Marketing Research

Analysis and Interpretation Results

Tests of between-subjects effects

Tests of Between-Subjects Effects

Dependent Variable: Income after the program

Source	Type III Sum of Squares	df	Mean Square	F	Sig.	Partial Eta Squared
Corrected Model	12420.421a	17	730.613	76.546	.000	.570
Intercept	101195.323	1	101195.323	10602.214	.000	.915
prog * incbef	12420.421	17	730.613	76.546	.000	.570
Error	9372.930	982	9.545			
Total	297121.000	1000				
Corrected Total	21793.351	999				

a. R Squared = .570 (Adjusted R Squared = .562)

It is seen from the output above the significance level of an interaction factor prog * incbef is less than 0.05; it indicates that the interaction is significant. The Partial Eta Square value of 0.570 indicates the amount of variation of the dependent variable of *Income after the program* which can be explained by the interaction of these two factors. The analysis will proceed in more depth with the following steps:

How to Conduct Analysis

- Please re-enable the box Univariate GLM dialog. Select **Analyze > General Linear Model > Univariate**
- Click **Model**
- choose **Full factorial** as the model type
- Click **Continue**
- Click **Options** GLM dialog box.
- choose **Descriptive statistics, Homogeneity tests, Spread vs. level plot,** and **Parameter estimates** in the **Display group**
- Click **Continue**
- Click **OK**

Analysis and Interpretation Results

Part I: Descriptive Statistics

Descriptive Statistics

Dependent Variable: Income after the program

Program status	Mean	Std. Deviation	N
0	14.4023	3.89303	517
1	18.9379	4.28162	483
Total	16.5930	4.67067	1000

The above output shows the average of income before the program and after the program. The average income before joining the program amounted to 14.4023 and after joining the program of 18.9379.

Section II: Levene's Test of Equality of Error Variances
Levene's Test of Equality of Error Variances[a]

Dependent Variable: Income after the program

F	df1	df2	Sig.
4.873	1	998	.028

Tests the null hypothesis that the error variance of the dependent variable is equal across groups.

a. Design: Intercept + incbef + prog

When viewed from the significance value in the above Sig column of 0.028 is smaller than 0.05; then it shows that there is a variance difference which is a violation of assumptions. However since there are only 2 cells defined by a combination of factor levels, this violation is less disturbing.

Part III: Spread-versus-level plot

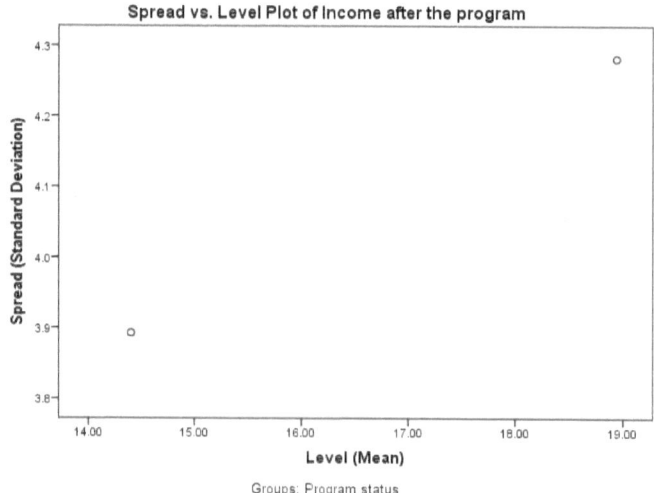

Groups: Program status

Data Analysis and Its Interpretation: Application in Marketing Research

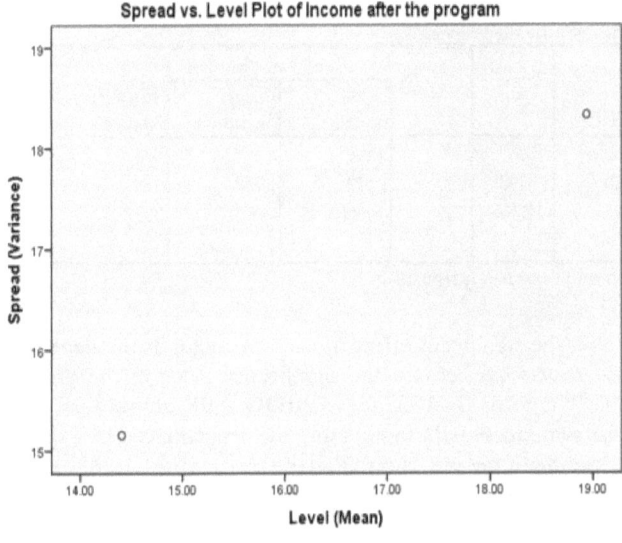

Spread-versus-level graph shows the relationship between the mean and standard deviation because only two combinations, then this cannot be used as a conclusion that a violation occurred. The difference in the distribution of only about 0.4 is small in comparison with the rate difference of 4.5. The conclusion is that the variance equality can also occur in this model.

Part IV: Between-subjects effects

Tests of Between-Subjects Effects
Dependent Variable: Income after the program

Source	Type III Sum of Squares	Df	Mean Square	F	Sig.	Partial Eta Squared
Corrected Model	12290.741a	2	6145.370	644.763	.000	.564
Intercept	131.400	1	131.400	13.786	.000	.014
Incbef	7153.844	1	7153.844	750.571	.000	.429
Prog	4735.662	1	4735.662	496.859	.000	.333
Error	9502.610	997	9.531			
Total	297121.000	1000				
Corrected Total	21793.351	999				

a. R Squared = .564 (Adjusted R Squared = .563)

The significance value of the participation and i in 0.000 < 0.05, indicating that the participation of the program have the significant effect on income. The R square values as much as 0.564 indicates the amount of variability of the income variable that can be explained by the participation of the program variable.

Section V : Parameter Estimation

Parameter Estimates

Dependent Variable: Income after the program

Parameter	B	Std. Error	T	Sig.	95% Confidence Interval		Partial Eta Squared
					Lower Bound	Upper Bound	
Intercept	4.197	.556	7.548	.000	3.106	5.288	.054
Incbef	1.636	.060	27.397	.000	1.519	1.753	.429
[prog=0]	-4.357	.195	-22.290	.000	-4.741	-3.974	.333
[prog=1]	0ª						

a. This parameter is set to zero because it is redundant.

Parameter estimates show the significant effect of each predictor in the dependent variable of *Income after the program* because the significance value in the program column is 0.000 < 0.05. The value is -4.357 for **[PROG = 0]** shows that, there are two people with the same income before joining the program, can be expected after the program the income of people who follow the program will be $ 4,357 and people who do not follow the program will be reduced by $ 4,357.

Summary

By specifying an interaction between the covariate and the factor, we can examine the similarity of covariate parameter estimation at all factor levels. Analysis of covariance shows that participation and the program increase salaries average of $ 4.357. A significant interaction effect shows that the assessment program participation becomes increasingly complex with the interaction between the factors of income and program participation before the program. When the interaction is significant, the difference between participating and non-participating people is changing due to the difference in income.

6.8 Using Univariate GLM For "Random Effects"

In this section we will add a store location variable as a *random effect* in addition to the previous review of the relationship between spending behavior and the amount of money spent. By adding the store location as the random effect, we can reduce the amount of unexplained variations so that it will improve accuracy in estimating the model. The file to be used is grocery_1month.sav.

How to conduct the analysis is as follows.

- Analyze > General Linear Model > Univariate
- Click **Reset** first to return to the original view.
- choose *Amount spent* as a dependent variable.
- choose *Who shopping for* and *Use coupons* as fixed factors
  Click **Options**
- choose **Estimates of effect size**.
- Click **Continue**
- Click **OK**

Data Analysis and Its Interpretation: Application in Marketing Research

Tests of between-subjects effects

Tests of Between-Subjects Effects

Dependent Variable: Amount spent

Source	Type III Sum of Squares	df	Mean Square	F	Sig.	Partial Eta Squared
Corrected Model	1867084.001[a]	11	169734.909	37.796	.000	.551
Intercept	54651422.013	1	54651422.013	12169.668	.000	.973
shopfor	1329509.066	2	664754.533	148.026	.000	.466
usecoup	317508.903	3	105836.301	23.567	.000	.173
shopfor * usecoup	192031.603	6	32005.267	7.127	.000	.112
Error	1522377.820	339	4490.790			
Total	59475118.440	351				
Corrected Total	3389461.820	350				

a. R Squared = .551 (Adjusted R Squared = .536)

The test results show the effect of fixed factor of shopfor and usecoup having significance value of 0.000 < 0.05; This suggests two factors can affect significantly. Now we will try to add a variable of *Store ID* as a random factor that will change the model we create.

Adding a Random Effect

- Re-enable Univariate GLM   dialog box: **Analyze > General Linear Model > Univariate**
- choose *Store ID* as a random factor
- Click **Model**
- choose **Custom** as model type
- Select variables of *shopfor* and *usecoup* in the list box of **Factors and Covariates**
- choose **Main effects** from the **Build Term (s)** menu and select it **main effects** for the model.
- Block *shopfor* and *usecoup* in the list box of **Factors and Covariates**
- choose **Interaction** from **Build   Term   (s)** and select the **interaction** for the model.
- choose *storeid* on **Factors and Covariates** and insert it into the model.
- Click **Continue**.
- Click **OK**

Tests Between Subjects Effects

Tests of Between-Subjects Effects

Dependent Variable: Amount spent

Source		Type III Sum of Squares	df	Mean Square	F	Sig.	Partial Eta Squared
Intercept	Hypothesis	51402962.479	1	51402962.479	7173.147	.000	.991
	Error	479937.374	66.974	7166.027a			
Shopfor	Hypothesis	1109546.714	2	554773.357	144.646	.000	.508
	Error	1073908.578	280	3835.388b			
Usecoup	Hypothesis	253850.608	3	84616.869	22.062	.000	.191
	Error	1073908.578	280	3835.388b			
shopfor * usecoup	Hypothesis	138871.544	6	23145.257	6.035	.000	.115
	Error	1073908.578	280	3835.388b			
Storeid	Hypothesis	448469.242	59	7601.174	1.982	.000	.295
	Error	1073908.578	280	3835.388b			

a. .884 MS(storeid) + .116 MS(Error)
b. MS(Error)

By added random factor of *Store ID*, it reduces unexplained variability in the model marked with Error, from 1522377.820 (first process before any random factor) become 1073908.578 (second process after added random factor). Also it raises the value of Partial eta square on the Shopfor variable from 0.466 to 0.508 and Usecoup from 0.173 to 0.191 and the interaction factor Shopfor * Usecoup from 0.112 to 0.115. The conclusion is that the model gets better when we add a random factor other than a fixed factor.

Summary

In the above case, by adding a random factor, the model improves with reduced variability that cannot be explained in the model. Although random factors are sometimes less appropriate to be used to explain the variability of dependent variables; however, if the selection is appropriate then the random factor will add accuracy in reducing unexplained variability in the model.

6.9 Multivariate General Linear Model

To test the hypothesis relating to parameter estimation then MGLM employs the following assumptions:

- Errors are independent on each observation and on each independent variable in the model. This assumption should not be violated to produce a correct result.
- Covariance of dependent variable is constant in all cells. This issue becomes important when the number of observations differs in each combination of factor levels.
- On any dependent variable, errors have a normal multivariate distribution with an average value of 0 and

standard deviation of 1.

6.9.1 A Case Analyzed by Using MGLM Procedure

In this example we will use the IBM SPSS file with the file name of patlos_sample.sav, which contains the sample treatment records of patients receiving thrombolytics during the treatment of heart attack (MI). The purpose of this study is to look at the effect of thrombolytics on heart disease patients. If these drugs are effective, they (patients) will stay in the hospital for a short period of time which should be longer so patients will save on hospitalization costs.

To conduct the analysis, use the following steps

- **Analyze > General Linear Model > Multivariate ....** The dialog box will appear as follows:

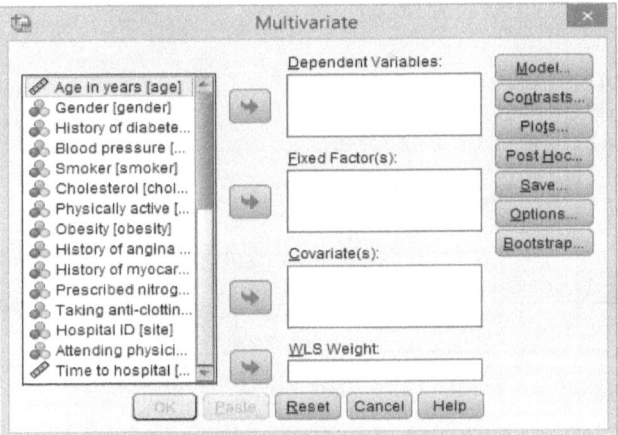

- choose *Length of stay* and *Treatment costs* dependent variables
- choose *Clot-dissolving drugs* and *Surgical treatment* as fixed factors
- Click **Contrasts**
- choose *clotsolv (None)* as contrast to change
- In groups of **Change Contrast group**, select **Simple** as contrast type
- choose **First** as **the reference category**
- Click **Change**, then **Continue**
- Click **Options** in the MGLM dialog box
- On dialog box of **Options**, select **Estimates of effect size**, **SSCP matrices**, **Homogeneity tests**, and **Spread vs. level plots** in the *Display group*
- Click **Continue**
- Click **OK**

The result and interpretation will be as follows.
Between Subjects Factors
The between subjects factors test result is as follows.

Between-Subjects Factors

		Value Label	N
Surgical treatment	1	PTCA	907
	2	CABG	574
Clot-dissolving drugs	1	Streptokinase	116
	2	Reteplase	696
	3	Alteplase	669

The between subjetcs factors section shows two treatments on the operation: PTCA (1) with a sample size of 907 and a CABG (2) with a sample size of 574. While for the use of Clot dissolving drugs, there are three, namely: Streptokinas (1) with a sample size of 116, Reteplase (2) with a sample size of 696 and Alteplase (3) with a sample size of 669.

Test of Between Subject Effects
The between subject effect test result is as follows.

Tests of Between-Subjects Effects

Source	Dependent Variable	Type III Sum of Squares	df	Mean Square	F	Sig.	Partial Eta Squared
Corrected Model	Length of stay	1217.307^a	5	243.461	202.406	.000	.407
	Treatment costs	79811.122^b	5	15962.224	865.665	.000	.746
Intercept	Length of stay	25234.532	1	25234.532	20979.169	.000	.934
	Treatment costs	1027759.201	1	1027759.201	55737.565	.000	.974
proc	Length of stay	727.562	1	727.562	604.872	.000	.291
	Treatment costs	44593.620	1	44593.620	2418.407	.000	.621
clotsolv	Length of stay	26.650	2	13.325	11.078	.000	.015
	Treatment costs	50.127	2	25.063	1.359	.257	.002
proc * clotsolv	Length of stay	6.757	2	3.379	2.809	.061	.004
	Treatment costs	50.182	2	25.091	1.361	.257	.002
Error	Length of stay	1774.185	1475	1.203			
	Treatment costs	27197.902	1475	18.439			
Total	Length of stay	47424.000	1481				
	Treatment costs	1970083.194	1481				
Corrected Total	Length of stay	2991.492	1480				
	Treatment costs	107009.024	1480				

a. R Squared = .407 (Adjusted R Squared = .405)
b. R Squared = .746 (Adjusted R Squared = .745)

The above output has the following meanings:
- The model that we created by using a dependent variable of Treatment Cost with fixed factors of Clot Disolving drugs and Surgical treatment is correct because the value of significance on both variables in the Corrected Model column is 0.000 < 0.05. The model is correct if the calculated significance value is less than 0.05.
- In Fixed Factor proc (Surgical Treatment) shows Length of stay and Treatment cost is significant because it has a significance value of 0.000 <

0.05. While in clotsolv only variable Length of stay is significant with value 0.000 < 0.05; vice versa for Treatment cost is not significant because the significance value of 0.257 > 0.05.
- The interaction between proc and clotsolv is not significant because the value of significance at Length of stay of 0.061> 0.05 and for Treatment cost of 0.257> 0.05.
- The value of R square (R^2) for the first part (proc) amounted to 0.407. This value has the significance of variable Colt Disolving drugs to Treatment cost variables of 0.407 or 40.7%.
- The value of R square (R^2) for the second part (clotsolv) of 0.746. This value means the effect of Surgical treatment on Treatment cost which is equal to 0.746 or 74.6%.

Multivariate Test Results Section
The multivariate test result is as follows.

Multivariate Test Results

	Value	F	Hypothesis df	Error df	Sig.	Partial Eta Squared
Pillai's trace	.026	9.833	4.000	2950.000	.000	.013
Wilks' lambda	.974	9.892a	4.000	2948.000	.000	.013
Hotelling's trace	.027	9.952	4.000	2946.000	.000	.013
Roy's largest root	.027	19.909b	2.000	1475.000	.000	.026

a. Exact statistic
b. The statistic is an upper bound on F that yields a lower bound on the significance level.

Multivariate testing using Pillai's trace, Wilks' lambda, Hotelling's trace and Roy's largest root shows a significance value of 0.000 which less than 0.05. Thus the model is significant.

Univariate Test Results Section
The univariate test result is as follows.

Univariate Test Results

Source	Dependent Variable	Sum of Squares	df	Mean Square	F	Sig.	Partial Eta Squared
Contrast	Length of stay	26.650	2	13.325	11.078	.000	.015
	Treatment costs	50.127	2	25.063	1.359	.257	.002
Error	Length of stay	1774.185	1475	1.203			
	Treatment costs	27197.902	1475	18.439			

Univariate test results show the Length of Stay variable has a significance value of 0.000 which is smaller than 0.05; thus the variables used as Fixed Factor: proc and clotsolv have a significant impact on the variable of Length of stay. Conversely not significant to Treatment cost variables because the value of significance of 0.257 greater than 0.05.

Between SSCP Matrix
The values of SSCP matrix can be seen the following table.

Between-Subjects SSCP Matrix

			Length of stay	Treatment costs
Hypothesis	Intercept	Length of stay	25234.532	161043.541
		Treatment costs	161043.541	1027759.201
	Proc	Length of stay	727.562	5696.019
		Treatment costs	5696.019	44593.620
	Clotsolv	Length of stay	26.650	-36.129
		Treatment costs	-36.129	50.127
	proc * clotsolv	Length of stay	6.757	18.413
		Treatment costs	18.413	50.182
Error		Length of stay	1774.185	3127.427
		Treatment costs	3127.427	27197.902

Based on Type III Sum of Squares

The table above shows the hypotheses as well as the sum-of-squares and cross-products (SSCP) sums squared matrix for model effect tests. Since there are two dependent variables, each matrix has two columns and two rows. For example the 2 x 2 matrix associated with CLOTSOLV is a matrix for testing the significance of Clot-dissolving drugs. Matrix associated with PROC is a hypothetical hypotesis matrix to test the significance of *Surgical treatment*, and the matrix associated with PROC * CLOTSOLV used to test the interaction effect. While error matrix is used to test each effect.

Multivariate Testing
The multivariate testing result can be seen in the following table

Data Analysis and Its Interpretation: Application in Marketing Research

Multivariate Tests[a]

Effect		Value	F	Hypothesis df	Error df	Sig.	Partial Eta Squared
Intercept	Pillai's Trace	.975	28781.280[b]	2.000	1474.000	.000	.975
	Wilks' Lambda	.025	28781.280[b]	2.000	1474.000	.000	.975
	Hotelling's Trace	39.052	28781.280[b]	2.000	1474.000	.000	.975
	Roy's Largest Root	39.052	28781.280[b]	2.000	1474.000	.000	.975
Proc	Pillai's Trace	.622	1212.157[b]	2.000	1474.000	.000	.622
	Wilks' Lambda	.378	1212.157[b]	2.000	1474.000	.000	.622
	Hotelling's Trace	1.645	1212.157[b]	2.000	1474.000	.000	.622
	Roy's Largest Root	1.645	1212.157[b]	2.000	1474.000	.000	.622
Clotsolv	Pillai's Trace	.026	9.833	4.000	2950.000	.000	.013
	Wilks' Lambda	.974	9.892[b]	4.000	2948.000	.000	.013
	Hotelling's Trace	.027	9.952	4.000	2946.000	.000	.013
	Roy's Largest Root	.027	19.909[c]	2.000	1475.000	.000	.026
proc * clotsolv	Pillai's Trace	.004	1.508	4.000	2950.000	.197	.002
	Wilks' Lambda	.996	1.508[b]	4.000	2948.000	.197	.002
	Hotelling's Trace	.004	1.509	4.000	2946.000	.197	.002
	Roy's Largest Root	.004	3.022[c]	2.000	1475.000	.049	.004

a. Design: Intercept + proc + clotsolv + proc * clotsolv
b. Exact statistic
c. The statistic is an upper bound on F that yields a lower bound on the significance level.

The multivariate test table displays four significance tests for each model effect:

- Pillai's trace is a positive statistical value in which the statistical values increasing shows an effect that contributes more to the model
- Wilks' Lambda is a positive statistical value which ranges from 0 to 1. The smaller value indicates the effect of contributing more to the model.
- Hotelling's trace is the sum of the Eigenvalues Matrix testing which is positive statistical value when the value increasing shows an effect which contributes more to the model. The value of Hotelling's trace is always greater than Pillai's trace; even when the value of Eigenvalues matrix testing is small; then both values will be the same. This indicates that the effect does not contribute to the model.
- Roy's largest root is the largest Eigenvalue test matrix value. This value is a statistical positive value where the value increasing shows an effect that makes more contribution to the model. The value of Roy's largest root is always smaller or equal to the value of Hotelling's trace. When both values are equal, the effect is dominated by one dependent variable, and there is a strong correlation between the dependent variables; and has a meaning effect does not contribute to the model significantly.

There is evidence that the value of Pillai's trace is more *robust* rather than statistical values others against violations of model assumptions. Each multivariate statistical value is transformed into the statistical value of testing close to the F value distribution. Degrees of freedom hypothesis (numerator) and error (denominator) for the F distribution will be displayed.

At the output above the values of significance of the main effect for the variables of *Clotsolv* and *Proc*, amounting to 0.000 which is less than 0.05; which shows the significant effect in the model. While the interaction effect between *Clotsolv* and *Proc* has a significance value of 0.197 which is greater than 0.05 which proves that the interaction effect has no significant effect on the model. The Partial eta squared statistic value shows the significance of *"practical"* for each part based on ratio of variations caused by the effect of the amount of variation caused by the remaining effect and variation in the error. More specifically, partial eta squared is a matrix product of SSCP matrix hypothesis and inversion amount matrix hypothesis and SSCP error. The greater the value of partial eta squared shows the greater the number of variations formed by the model effect with a maximum value of 1. The partial eta squared value for the variable of *Clotsolv* is very small, that is equal to 0.013, indicating that the variable does not contribute a significant effect to model. Otherwise the value of partial eta squared for the variable of *Proc* is 0.622, showing the significant effect on the model as expected. The conclusion is that the surgical procedure that a patient must undergo for *MI treatment* will have a greater effect on the length of stay in hospital and the total treatment cost compared to the type of thrombolytic they receive.

Contrast Results Section
The contrast result is as follows.

Data Analysis and Its Interpretation: Application in Marketing Research

Contrast Results (K Matrix)

Clot-dissolving drugs Simple Contrast[a]		Dependent Variable	
		Length of stay	Treatment costs
Level 2 vs. Level 1	Contrast Estimate	-.382	.593
	Hypothesized Value	0	0
	Difference (Estimate - Hypothesized)	-.382	.593
	Std. Error	.112	.439
	Sig.	.001	.176
	95% Confidence Interval for Difference Lower Bound	-.602	-.267
	Upper Bound	-.162	1.453
Level 3 vs. Level 1	Contrast Estimate	-.516	.722
	Hypothesized Value	0	0
	Difference (Estimate - Hypothesized)	-.516	.722
	Std. Error	.112	.439
	Sig.	.000	.100
	95% Confidence Interval for Difference Lower Bound	-.736	-.138
	Upper Bound	-.296	1.583

a. Reference category = 1

The output table above shows the results for each contrast in the model. Simple contrast using first level *Clot-dissolving drugs* as a reference category. Thus, one contrast is done by comparing the second level with first level, i.e. an effect of *reteplase* with effect *streptokinase*. Contrast estimates show average of the patients given *reteplase* spent 0.382 days shorter in the hospital and spend 600 dollars more to the cost of treatment compared with patients who are given *streptokinase*. Because the value of significance for *Length of stay* of 0.001 is smaller than 0.05, it can be concluded that this difference is significant. While the value for significance of *Treatment costs* as much as 0.176 greater than 0.05, then this difference is insignificant.

The second contrast compares the third level with the first level, i.e. the effect of *alteplase* with effect of *streptokinase*. The contrast estimates show the average of patients given *alteplase* spent half a day less in the hospital and spent $ 700 more on maintenance. Because of the significance value of *Length of stay* as much as 0.000 is smaller than 0.05, then the difference is really significant. While the significance value of *Treatment costs* as much as 0.10 is greater than 0.0 5, then the difference is not significant. In conclusion the contrast shows that *alteplase* d an *reteplase* have reduced the length of a patient's stay in the hospital. This reduction is sufficient to meet the cost of maintenance. Thus the model shows that *alteplase* and *reteplase* must be used to replace *streptokinase* .

Covariance Matrices Equality Test Using Box's M

The result of the covariance matrices equality test using Box's M is as follows.

Box's Test of Equality of Covariance Matrices[a]

Box's M	270.509
F	17.908
df1	15
df2	358296.484
Sig.	.000

Tests the null hypothesis that the observed covariance matrices of the dependent variables are equal across groups.

a. Design: Intercept + proc + clotsolv + proc * clotsolv

The assumption for a multivariate procedure is the vector of dependent variables follows the normal distribution and the same covariance matrix for all cells formed by the effect between the subjects. Box's M is used to perform the test of the null hypothesis in which the covariance matrix of the dependent variable is observed equally in all groups. This value is transformed into F statistical value with Degree of Freedom (DF) 1 and DF 2. In this case Box's M value of 0.000 smaller than 0.05 indicates that there is a violation of the assumption of equality of the covariance matrix. The value of the covariance matrix is said to be the same if the calculated significance value is greater than 0.05.

The above output yields a significance value of 0.000 which is smaller than 0.05. Thus there is a violation of the assumption that the value of significance should be greater than 0.05. This can be explained by performing hypothesis testing as follows:

H0: The covariance matrix of the dependent variable is observed equally in all groups
H1: The covarianed matrix of the dependent variable observed is not the same across all groups

Hypothesis Testing Criteria
If the significance value < 0.05; then H0 is rejected and H1 accepted
If the significance value > 0.05; then H0 is accepted and H1 is rejected

Decision
Because the significance value as much as 0.000 < 0.05; then H0 is rejected and H1 accepted. Thus the covariance matrix of the dependent variable observed is not the same across all groups.

Levene's Test of Error Variance Equality
The following is the result of the test of error variance equality.

Data Analysis and Its Interpretation: Application in Marketing Research

Levene's Test of Equality of Error Variances[a]

	F	df1	df2	Sig.
Length of stay	1.507	5	1475	.185
Treatment costs	10.001	5	1475	.000

Tests the null hypothesis that the error variance of the dependent variable is equal across groups.
a. Design: Intercept + proc + clotsolv + proc * clotsolv

Testing the variance equality can be done as follows:
First: A variable of length of stay
Hypothesis
H0: error variance of the dependent variable observed is equal in all groups
H1: error variance of the dependent variable observed is not equal in all groups

Hypothesis Testing Criteria
If the significance value < 0.05; then H0 is rejected and H1 accepted
If the significance value > 0.05; then H0 is accepted and H1 is rejected

Decision
Because the significance value as much as 0.185 > 0.05; then H0 is accepted and H1 is rejected. Thus, the error variance observed is equal in all groups. Thus in this first variable there is no violation of the assumption because the error variance for the dependent variable is equal.

Second: A variable of treatment costs
Hypothesis
H0: error variance of the dependent variable observed is equal in all groups
H1: error variance of the dependent variable observed is not equal in all groups

Hypothesis Testing Criteria
If the significance value < 0.05; then H0 is rejected and H1 accepted
If the significance value > 0.05; then H0 is accepted and H1 is rejected

Decision
Because the significance value as much as 0.000 < 0.05; then H0 is rejected and H1 accepted. Thus the error variance of the dependent variable observed is not equal in all groups. Accordingly in this second variable there is a violation of the assumption saying that the error variance of the dependent variable observed is equal in all groups.

Graph Section: Spread vs. level plot for Treatment costs

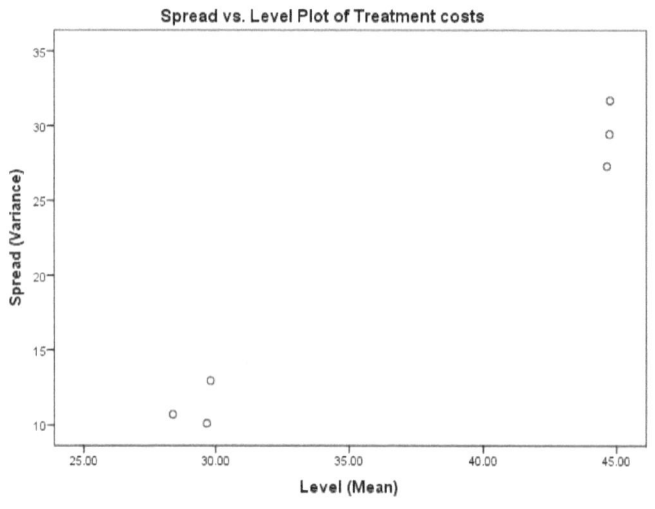

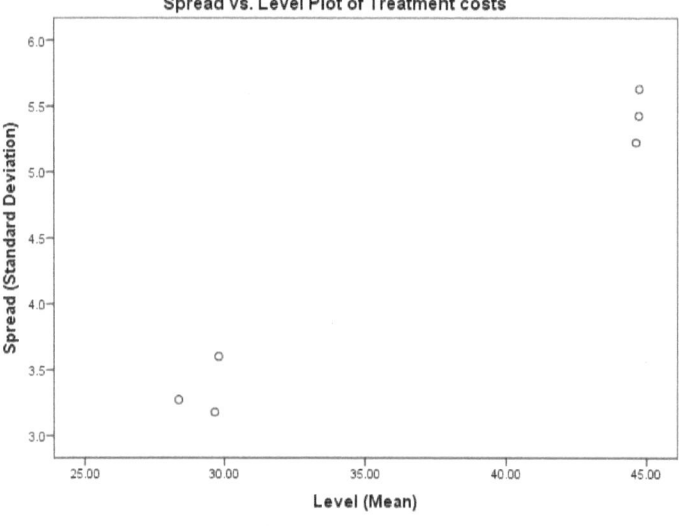

This graph is the average distribution chart and standard deviation that describes the visual assumption of variance equality test. In order to overcome the violation occuring in the GLM procedure, we will use the Log- cost variable instead of the Treatment cost variable.

Using the Logarithmic Transformable Treatment Cost
The steps of analysis are as follows:
1. Analyze> General Linear Model> Multivariate.
2. choose *Log-cost* as the dependent variable .
3. Click **OK**

Data Analysis and Its Interpretation: Application in Marketing Research

The result will be as below

Interpretation

Covariance Matrix Equality Testing
The following is the result of covariance matrix equality test.

Box's Test of Equality of Covariance Matrices[a]

Box's M	56.228
F	3.722
df1	15
df2	358296.484
Sig.	.000

Tests the null hypothesis that the observed covariance matrices of the dependent variables are equal across groups.
a. Design: Intercept + proc + clotsolv + proc * clotsolv

Box's M results are still the same as before, i.e. a significance value of 0.000 smaller than 0.05 indicating the inequality of error variance in the dependent variable still occurs.

Error Variance Equality Using Levene Testing
To test the error variance using Levene can also detect the equality of error variance among the dependent variables. The result if as follows

Levene's Test of Equality of Error Variances[a]

	F	df1	df2	Sig.
Length of stay	1.507	5	1475	.185
Log-cost	1.291	5	1475	.265

Tests the null hypothesis that the error variance of the dependent variable is equal across groups.
a. Design: Intercept + proc + clotsolv + proc * clotsolv

The Levene test shows that the second log-cost variable now has a significance value of 0.265 > 0.05 which, when using the Cost variable alone, has a significance value of 0.000. Thus Levene testing for the error variance equality has been met. While Box's M testing can not be fulfilled; on testing the similarity of error variance of both variables of Fixed Factors has already met the requirements of error variance equality. Thus the process of analysis can be continued.

Multivariate Testing
Now we come to the multivariate testing with result is as follows.

Multivariate Tests[a]

Effect		Value	F	Hypothesis df	Error df	Sig.	Partial Eta Squared
Intercept	Pillai's Trace	.998	482037.856[b]	2.000	1474.000	.000	.998
	Wilks' Lambda	.002	482037.856[b]	2.000	1474.000	.000	.998
	Hotelling's Trace	654.054	482037.856[b]	2.000	1474.000	.000	.998
	Roy's Largest Root	654.054	482037.856[b]	2.000	1474.000	.000	.998
proc	Pillai's Trace	.663	1447.653[b]	2.000	1474.000	.000	.663
	Wilks' Lambda	.337	1447.653[b]	2.000	1474.000	.000	.663
	Hotelling's Trace	1.964	1447.653[b]	2.000	1474.000	.000	.663
	Roy's Largest Root	1.964	1447.653[b]	2.000	1474.000	.000	.663
clotsolv	Pillai's Trace	.033	12.293	4.000	2950.000	.000	.016
	Wilks' Lambda	.967	12.387[b]	4.000	2948.000	.000	.017
	Hotelling's Trace	.034	12.481	4.000	2946.000	.000	.017
	Roy's Largest Root	.034	24.873[c]	2.000	1475.000	.000	.033
proc * clotsolv	Pillai's Trace	.005	1.774	4.000	2950.000	.131	.002
	Wilks' Lambda	.995	1.775[b]	4.000	2948.000	.131	.002
	Hotelling's Trace	.005	1.776	4.000	2946.000	.131	.002
	Roy's Largest Root	.005	3.554[c]	2.000	1475.000	.029	.005

a. Design: Intercept + proc + clotsolv + proc * clotsolv
b. Exact statistic
c. The statistic is an upper bound on F that yields a lower bound on the significance level.

In multivariate testing using Pillai's trace, Wilks' lambda, Hotelling's trace and Roy's largest root on the proc and clotsolv variables shows a significance value of 0.000 smaller than 0.05. Thus multivariate test is significant.

Test of Between Subjects Effects

The testing of between subject effects results in the following table.

Tests of Between-Subjects Effects

Source	Dependent Variable	Type III Sum of Squares	df	Mean Square	F	Sig.	Partial Eta Squared
Corrected Model	Length of stay	1217.307a	5	243.461	202.406	.000	.407
	Log-cost	59.379b	5	11.876	1022.151	.000	.776
Intercept	Length of stay	25234.532	1	25234.532	20979.169	.000	.934
	Log-cost	9638.025	1	9638.025	829543.822	.000	.998
Proc	Length of stay	727.562	1	727.562	604.872	.000	.291
	Log-cost	33.552	1	33.552	2887.814	.000	.662
Clotsolv	Length of stay	26.650	2	13.325	11.078	.000	.015
	Log-cost	.057	2	.028	2.443	.087	.003
proc * clotsolv	Length of stay	6.757	2	3.379	2.809	.061	.004
	Log-cost	.059	2	.030	2.541	.079	.003
Error	Length of stay	1774.185	1475	1.203			
	Log-cost	17.137	1475	.012			
Total	Length of stay	47424.000	1481				
	Log-cost	18656.492	1481				
Corrected Total	Length of stay	2991.492	1480				
	Log-cost	76.516	1480				

a. R Squared = .407 (Adjusted R Squared = .405)
b. R Squared = .776 (Adjusted R Squared = .775)

The above output has the following meanings:

- The model we created uses the dependent variable of Treatment Cost with the fixed factors of Clot Disolving drugs and Surgical treatment is correct because the value of significance on both variables in the Corrected Model column is 0.000 < 0.05. The model is correct if the calculated significance value is less than 0.05.
- In Fixed Factor of proc (Surgical Treatment) showing that Length of stay and Log-cost are significant because the significance value of 0.000 < 0.05. While in clotsolv only variable Length of stay is significant with value 0.000 < 0.05. Therefore, Log-cost is not significant because the significance value is 0.087> 0.05.
- The interaction between proc and clotsolv is not significant due to the significance value of Length of stay is 0.061 > 0.05 and for Log- cost of 0.079 > 0.05.
- The value of R square (R^2) for the first part (proc) amounted to 0.407. This value shows the significant effect of Colt Disolving drugs on the Treatment cost variable as much as 0.407 or 40.7%.
- The value of R square (R^2) for the second part (clotsolv) is 0.776. This value shows the significant effect of the Surgical Treatment variable on the Log-cost variable as much as 0.776 or 77.6 %.

Between Subjects SSCP Matrix

The values of between subjects SSCP matrix can be seen the table below.

Between-Subjects SSCP Matrix

			Length of stay	Log-cost
Hypothesis	Intercept	Length of stay	25234.532	15595.225
		Log-cost	15595.225	9638.025
	Proc	Length of stay	727.562	156.241
		Log-cost	156.241	33.552
	Clotsolv	Length of stay	26.650	-1.177
		Log-cost	-1.177	.057
	proc * clotsolv	Length of stay	6.757	.632
		Log-cost	.632	.059
Error		Length of stay	1774.185	88.403
		Log-cost	88.403	17.137

Based on Type III Sum of Squares

The table above shows the hypotheses as well as the sum-of-squares and cross-products (SSCP) matrix for model's effect tests. Since there are two dependent variables, each matrix has two columns and two rows. For example the 2 x 2 matrix associating with CLOTSOLV is a matrix for testing the significance of Clot-dissolving drugs. Matrix associating with PROC is a hypothetical hypothesis to test the significance of *Surgical treatment*, and the matrix associating with PROC * CLOTSOLV is used to test the interaction effect. While error matrix is used to test each effect.

Multivariate Test Results

The result of the multivariate can be seen the table below

Multivariate Test Results

	Value	F	Hypothesis df	Error df	Sig.	Partial Eta Squared
Pillai's trace	.033	12.293	4.000	2950.000	.000	.016
Wilks' lambda	.967	12.387[a]	4.000	2948.000	.000	.017
Hotelling's trace	.034	12.481	4.000	2946.000	.000	.017
Roy's largest root	.034	24.873[b]	2.000	1475.000	.000	.033

a. Exact statistic
b. The statistic is an upper bound on F that yields a lower bound on the significance level.

Multivariate tests for the overall model using Pillai's trace, Wilks' lambda, Hotelling's trace and Roy's largest root show a significance value of 0.000 which is smaller than 0.05. Thus the multivariate test of the model as a whole is significant.

Data Analysis and Its Interpretation: Application in Marketing Research

Univariate Test Results
The univariate test result are as follows

Univariate Test Results

Source	Dependent Variable	Sum of Squares	df	Mean Square	F	Sig.	Partial Eta Squared
Contrast	Length of stay	26.650	2	13.325	11.078	.000	.015
	Log-cost	.057	2	.028	2.443	.087	.003
Error	Length of stay	1774.185	1475	1.203			
	Log-cost	17.137	1475	.012			

Univariate test results show the Length of Stay variable has a significance value of 0.000 which is smaller than 0.05; thus the variables used as Fixed Factors: proc and clotsolv have a significant impact on the variable of Length of stay. Conversely not significant to the Log-cost variable because the significance value of 0.087 is greater than 0.05.

Contrast Section
The contrast result is seen below

Contrast Results (K Matrix)

Clot-dissolving drugs Simple Contrast[a]		Dependent Variable	
		Length of stay	Log-cost
Level 2 vs. Level 1	Contrast Estimate	-.382	.022
	Hypothesized Value	0	0
	Difference (Estimate - Hypothesized)	-.382	.022
	Std. Error	.112	.011
	Sig.	.001	.049
	95% Confidence Interval Lower Bound for Difference	-.602	.000
	Upper Bound	-.162	.043
Level 3 vs. Level 1	Contrast Estimate	-.516	.024
	Hypothesized Value	0	0
	Difference (Estimate - Hypothesized)	-.516	.024
	Std. Error	.112	.011
	Sig.	.000	.028
	95% Confidence Interval Lower Bound for Difference	-.736	.003
	Upper Bound	-.296	.046

a. Reference category = 1

Contrast to Level 1 versus Level 2 and Level 3 versus Level 1
The contrast for Level 1 versus Level 2 is significant because the significance value in the Length of Stay variable is 0.001 <0.05 and the Log-cost variable is 0.049 < 0.05. Similarly, in contrast to Level 3 and 1 the significance value in the Length of

Stay variable of 0.000 <0.05 and the Log-cost variable of 0.028 <0.05. The conclusion is that the contrast we make in the model is significant.

Summary

The initial model in this study shows that treatment of costs for reteplase and alteplase variables does not differ significantly from the cost of treatment with streptokinase. This happens because there is a violation of the error variance equality assumption. That is why we use the Log-cost variable instead of Treatment costs. After the replacement we see in the Levene test there is no violation of the assumption of error variance equality.

CHAPTER 7

HOW TO MAKE A GRAPH TO PRESENT RESEARCH RESULTS

7.1 How to Make a Communicative and Easy Graph

Graphs help a lot in presenting research data effectively if we can design well. In the following section, we will discuss how to create useful graphs to display the results communicatively.

To create a graph use the following ways:

- Select **Graphs > Chart Builder**

The dialog box will appear as follows.

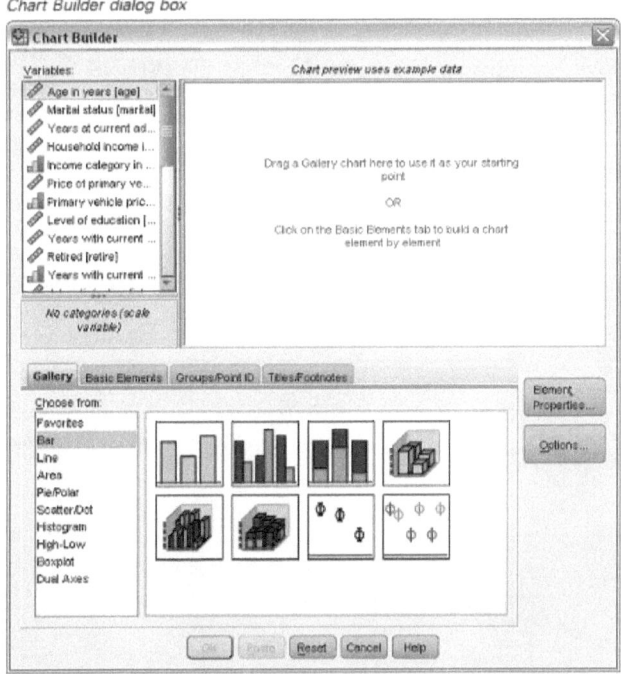

Chart Builder dialog box

- Select the Bar in the Gallery column

Bar chart on Chart Builder canvas

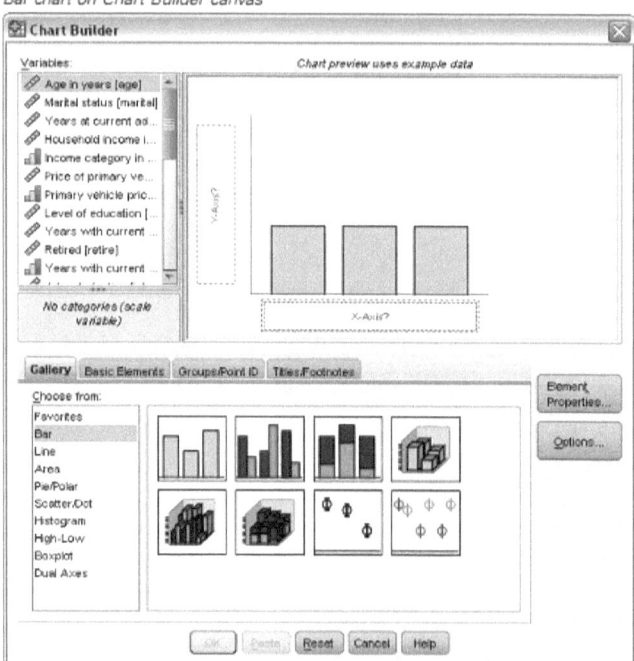

- Move the Job satisfaction variable from the Variables list to the x axis as shown below

Data Analysis and Its Interpretation: Application in Marketing Research

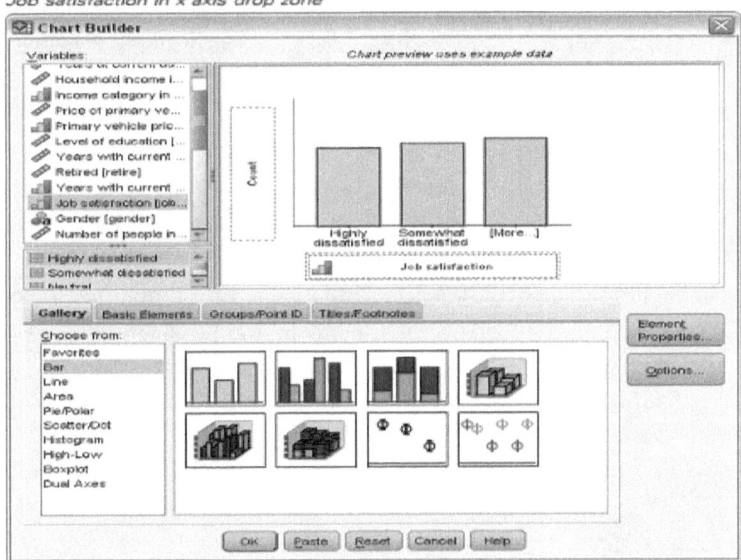

- K lik **Element Properties** to display the window **Element Properties** such as the following.

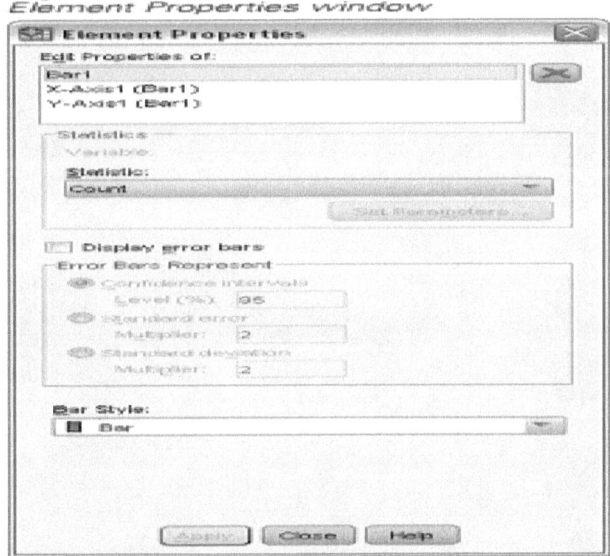

- Return to the Chart Builder dialog box
- Move the Household income in thousands variable from the Variables list to the y- axis. Since the variable on the y axis is scalar

(the real number (the value representing the quantity on the continuous line)) is on the x axis variab e l categorical al, then on the y axis will display the average value.

To add a title, do the next steps:
- Click **Titles / Footnotestab** .
- Select **Title 1**

In the **Properties** window, select **Title 1** in the **Edit Properties** section.
- In the **Content** column type **Income by Job Satisfaction** .
- Click **Apply** to save the text. Although the text does not appear on Chart Builder, but the text will appear when we activate the graph.

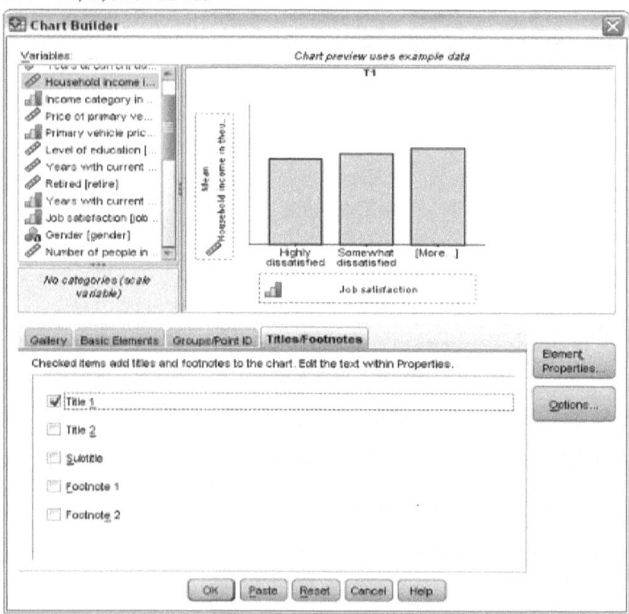

The title appears on the canvas with the label T1.

- Click OK to make **bar chart**

Data Analysis and Its Interpretation: Application in Marketing Research

Bar chart

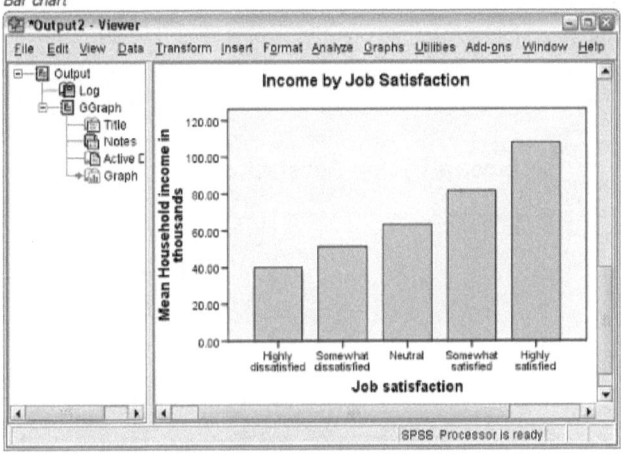

The bar chart reveals that respondents who are more satisfied with their jobs tend to have higher household incomes.

7.2 How to Perform Graphics Editing

We can do graph editing relating to the following things:
- Change the graph coloring
- Format numbers on labels
- Edit text
- Displays the data value label
- Use chart templates
- To edit graphics

To open Chart Editor, double click on the bar chart. Chart Editor will open as below.

Bar chart in the Chart Editor

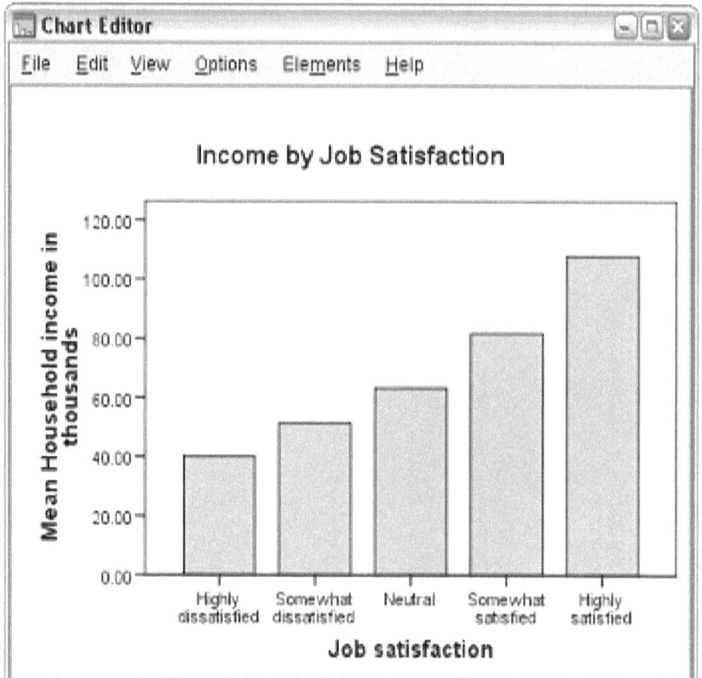

7.2.1 Selecting Graph Elements

To perform graphical element editing, the steps are as follows:
- Click part of one image graph until a box that surrounds the trunk.

There are general rules for selecting elements in simple charts:
- When no graphic elements are selected, click any graphic elements to select all the graphic elements.
- When all the graph elements are selected, click one graph element to select only the graph element. We can choose different graph elements by clicking on them. To select multiple graphic elements, click each of these elements while holding down the Ctrl key.
    - To delete the selection press Esc key.
    - Click any part of the bar to re-select the image of the bar.

7.2.2 Using Properties Window
From the Chart Editor menu choose :
Edit > Properties

Data Analysis and Its Interpretation: Application in Marketing Research

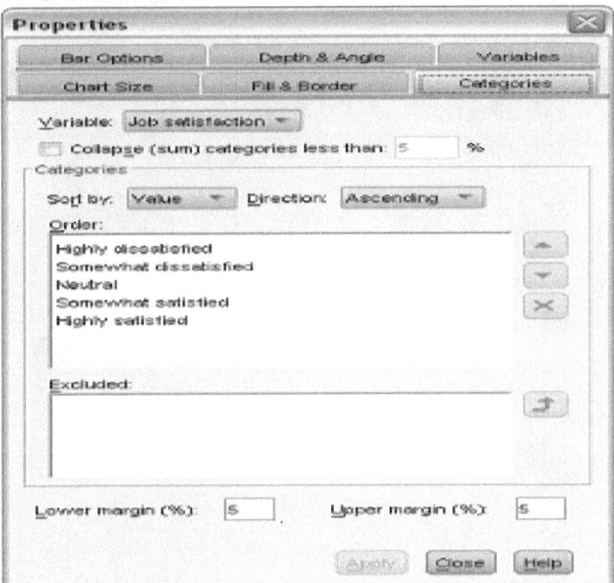

7.2.3 Changing the Barcode Color

The first time we will change the color of the stem image. Specify the color attribute specifics of the graph elements in the **Fill & Border** command.

- Click **Fill & Border.**
- Click color box after command **Fill** to indicate that we will change the color of the contents on the images of the bar. Number of colorful boxes to make speeches of red, green, and blue bluebackgrounds for current staining.
- Click light blue color on the second from the left on the second row at the bottom.
- Click **Apply** .

Fill & Border tab

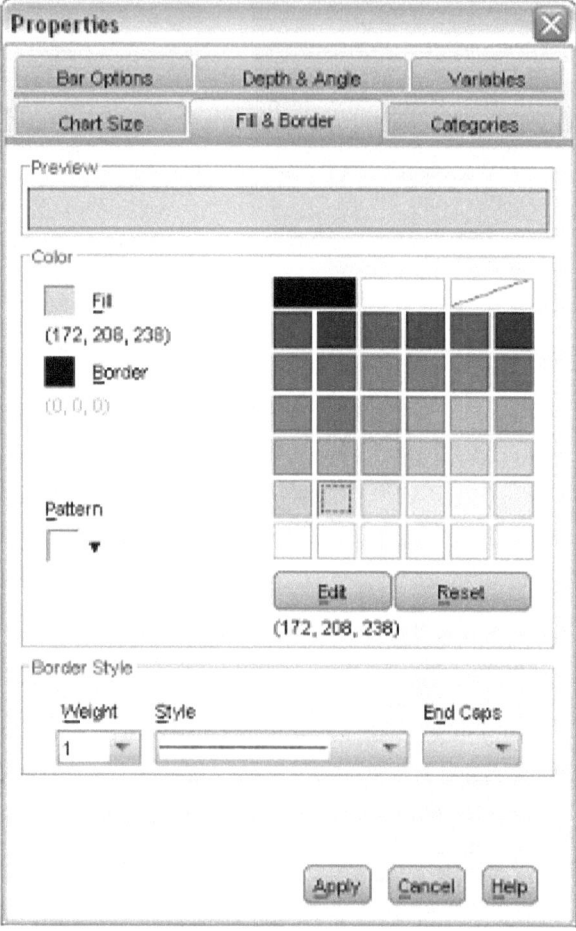

The results show the bars in the graph become light blue at this time.

7.2.4 Creating Number Format in Labels

Please note that the numbers on the y- axis are measured in thousands. In order for the graphics to be more interesting and easier to interpret, then we will change the number format in the label and then edit the title correctly.
- Select the y- axis on the label by clicking one of the images.
- To reopen the **Properties** window if we close again, the way is, choose : **Edit > Properties**
- Click **Number Format**.

- If we do not want the label to display decimal, then type 0 in the text box in **Decimal Places**
- Type 0.001 in the text box **Scaling Factor**. Then the measurement factor is where **Chart is Editor** dividing the number displayed. Since 0.001 is a fractional number, by dividing it then that number will increase on a label of 1.000. Thus the numbers are not in the form of thousands; the numbers will not be measured.
- Select **Display Digit Grouping**. **Digit** Grouping uses the default of each of our computers to mark the place of every thousand in that number.
- Click **Apply**

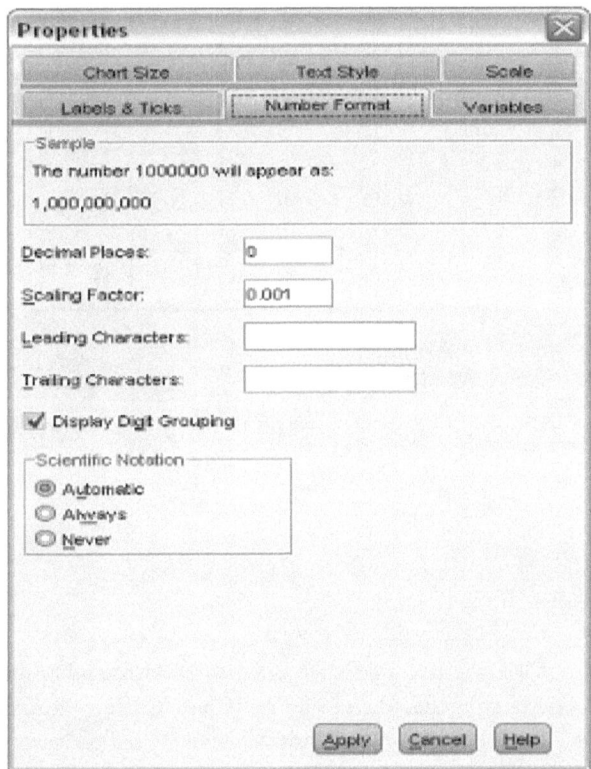

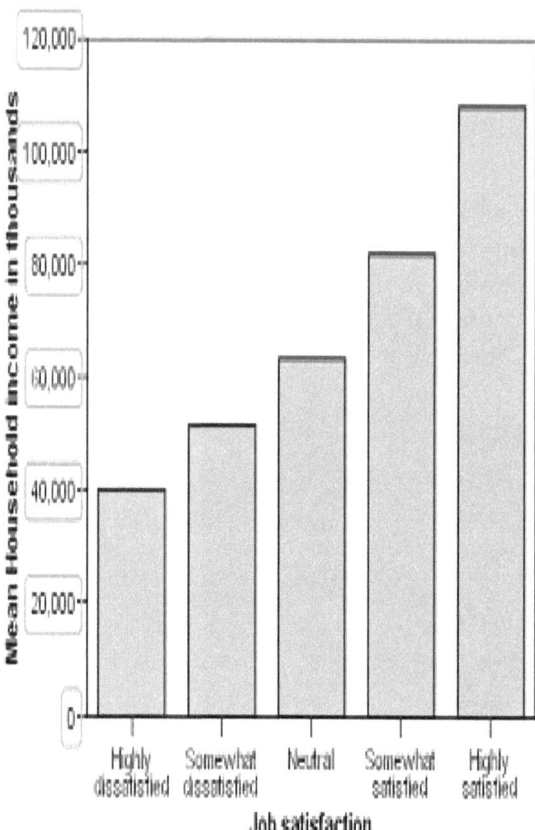

7.2.5 Editing Text
To do text editing how:
- Click a y axis title to select it .
- Click the axis title once again to start in editing mode. While in editing mode, the position of **Chart Editor** will rotate the text in horizontal. It will also display a blinking red bar cursor
- Clear text : *in thousands*
- Press **Enter** to exit editing mode and update axis title. So the current axis title accurately describes the contents of the label.

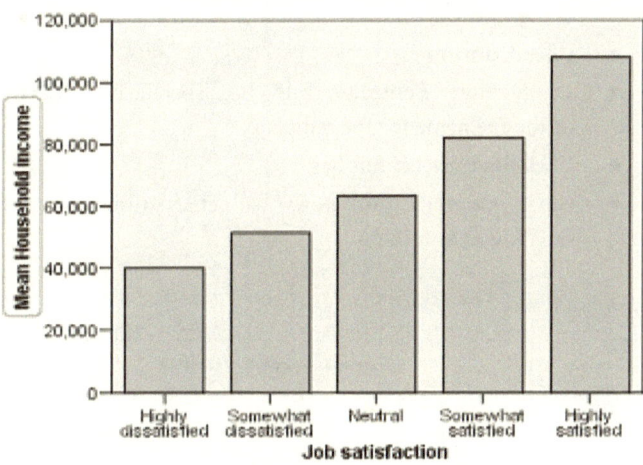

7.2.6 Display Data Value Label

To display the data value label we can do as follows:
From the Chart Editor menu select

- **Elements > Show Data Labels**

Then it will come out the view as follows:

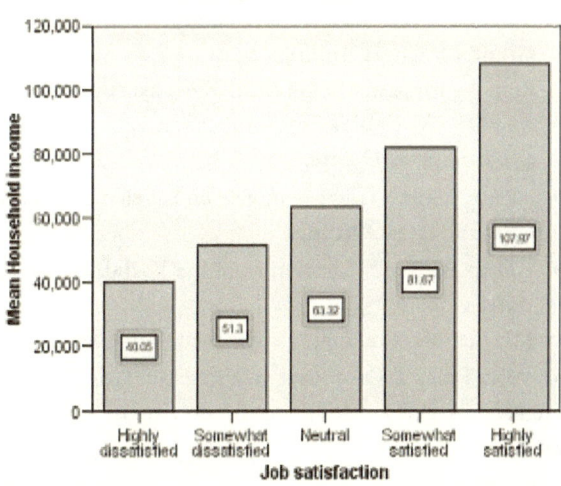

7.2.7 Using Template

Creating a chart with the template menu can be done as follows

- **File > Save Chart Template**

- Select all settings to include all available graphic attributes in the template
- Click **Continue**
- In the **Save Template** dialog box, specification of the location of the file name for the template.
- Click **Save** after it finishes
- Close **Chart Editor**. New bar chart results can be viewed on **Viewer** as follows.

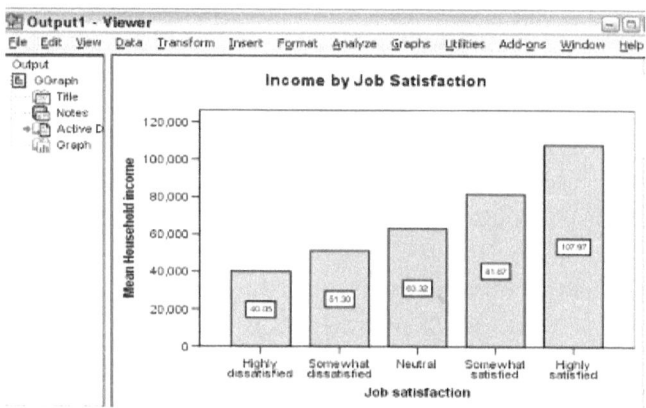

Then from the menu **Viewer** choose :
Graphs > Chart Builder
Remove Job satisfaction from the x-axis by pulling and laying in the area of **Variables** list. Another way is to click on the *drop zone* d an press **Delete**.

- Click the right mouse on Level of education on **Variables list** and select **Ordinal**
- Drag Level of education from the **Variables list** to the x- axis image area column

Currently the title is no longer appropriate for it we will delete it .
- On **Titles / Footnotes tab**, deselect **Title 1**

Currently tips will create a specification in the template to apply the new graph
- Click **Options**
- On **Templates group** in **Options** dialog box, click **Add**
- In the dialog box **Find Template Files**, place the template file which we have previously saved using the commands in the dialog box of

Save Chart Template

- Select the file and click **Open**

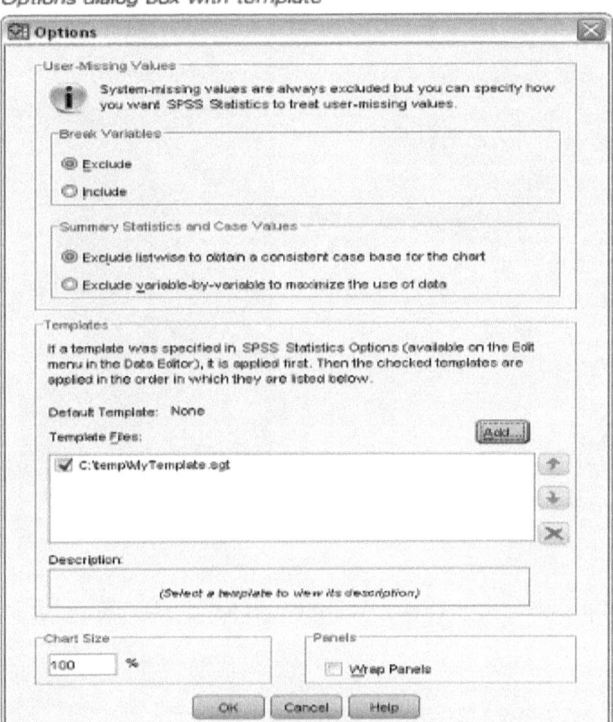

Options dialog box with template

Click **OK** to close the dialog box **Options**

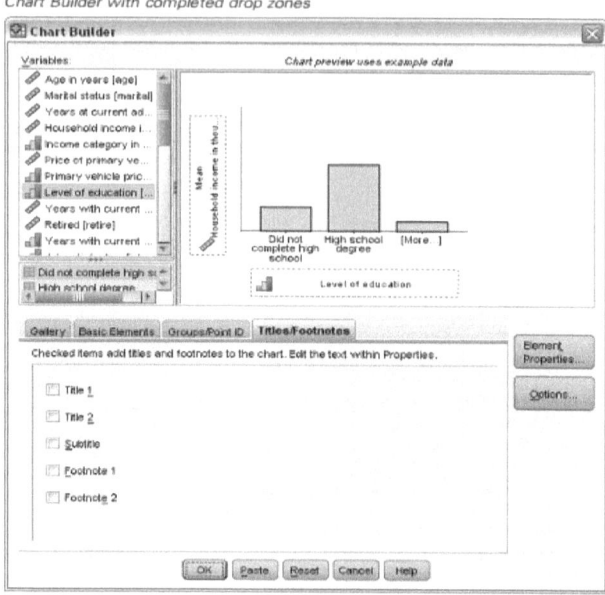

Click **OK** in the dialog box **Chart Builder** to make graph and apply the template.

7.2.8 Defining Options in Graph

From the Data Editor or Viewer menu select :

- **Edit > Options**

Data Analysis and Its Interpretation: Application in Marketing Research

Charts tab in Options dialog box

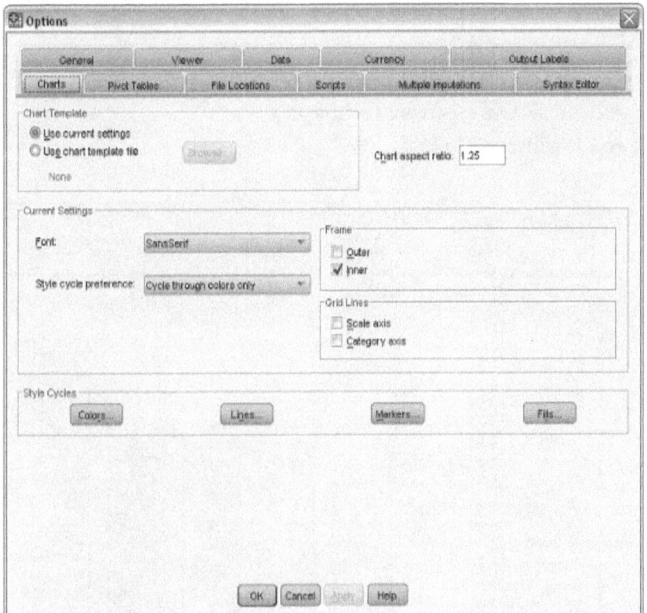

These options control how the graph will be created. For each new graph we can create a specification:
- Whether we will use the current settings or templates
- Width and height ratio
- If we do not use the template, seting to make the format of the cycle style of the graph elements.
- Click **Colors** to open a dialog box **Element Colors Data**
- Choose **Simple Charts**
- Choose a light green color
- Click **Continue**
- In the **Options** dialog box, click OK to save the color style cycle changes. Then the graph elements on each new chart will be light green.

From **Data Editor** or **Viewer** menu choose :
- **Graphs > Chart Builder**

Then Chart Builder will display the last graph we have created. Keep in mind that this graph is still associated with the template earlier. If we do not

want to use the template; then do:
- Click **Options**
- Remove option on that template
- Click **OK** to create the graph.

The result will be as below.

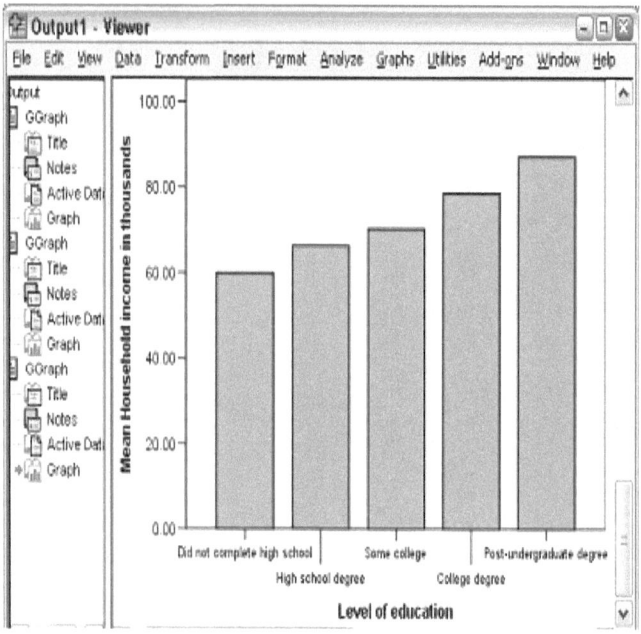

CHAPTER 8

A MARKETING RESEARCH EXAMPLE USING LINEAR REGRESSION CONSIDERING THE CLASSICAL ASSUMPTION

8.1 The First Case

A distributor of cars conducted a research with the aim is to know whether number of direct sales force affects the sale as well as to know whether the sale can be predicted by using the number of direct sales force? This study uses data for twelve months. The data is shown below:

Month	Sale	Direct Sales Force
January	200	30
February	140	25
March	130	15
April	175	24
May	240	31
June	120	27
July	160	10
August	90	9
September	300	29
October	89	12
November	70	16
December	60	14

The formulation of the problems are as follows:

• What is the effect of the direct sales force on cars' sale?

• What is the likely cars' sale in the future?

8.1.1 Problem Solution

If seen from the model relationship of the above variables, then we can identify as follows:

• The direct sales force is an independent variable and which is also called a predictor.

- The cars' sale is a dependent variable.

To solve this problem we will use a simple linear regression procedure that matches the problem in the study.

8.1.1.1 Basic Concepts of Linear Regression

Linear regression is an equation model based on a straight line reflecting the linear relationship between the independent variable (X) and the dependent variable (Y). The straight line can be defined by two things: first, the slope or gradient represented by the notation of β1 and second, a point where the line crosses the vertical axis in the graph, called the intercept and represented by the notation of a or β0. In this book the notation of "a" is used.

Thus we will create a model using our own data to be tailored to a straight line model that reflects the linear relationship between variable of labor force with variable of indirect production cost. To understand the linear regression correctly we will see the image of the regression line as below.

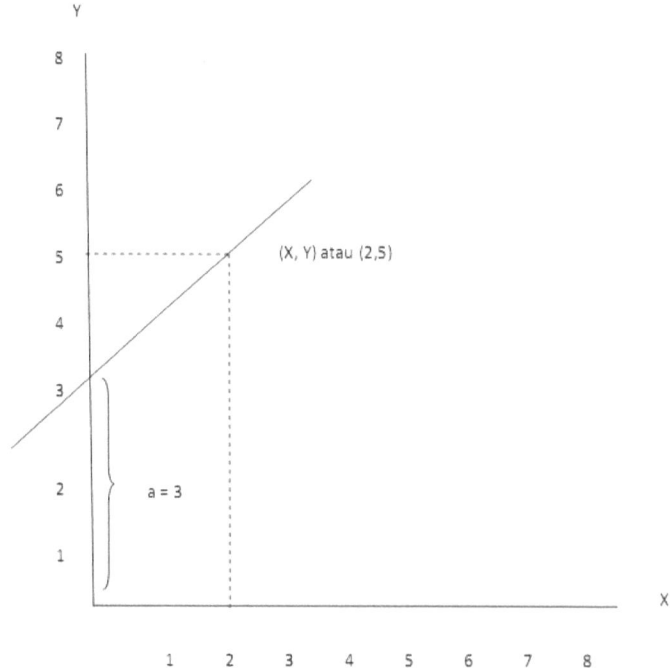

Figure 8.1 Regression Lines

The equation is

$$Y = a + \beta_1 X_1$$

With:

Y = dependent variable / criterion

a = intercept of Y (In IBM SPSS it is called constant)

β = slope (also called slope or gradient and in IBM SPSS it is called regression coefficient)

X = independent variable

8.1.1.2 Equations in Simple Linear Regression

Linear regression has an equation which is called the regression equation. The regression equation expresses the linear relationship between the dependent variable (criterion variable) given the Y symbol and one or more independent variables (predictors) given the X symbol if there is only one predictor and X1, X2 to Xk, if there are more than one predictor (Crammer & Howitt, 2006: 139). The regression equation should look like this:

For a regression equation where Y is a predicted value, the equation is:

$$Y = a + \beta_1 X_1 \text{ (for simple linear regression)}$$

$$Y = a + \beta_1 X_1 + \beta_2 X_2 + ... + \beta_k X_k \text{ (for multiple linear regression)}$$

For the regression equation where Y is the absolute value (observation), then the equation includes the error term (residual) will be:

$$Y = a + \beta_1 X_1 + e \text{ (for simple linear regression)}$$

$Y = a + \beta_1 X_1 + \beta_2 X_2 + ... + \beta_k X_k + e$ (for multiple linear regression)

Where:

- X: is the true value of a case (data) on the independent variable (X)

- β: is a regression coefficient if there is only one predictor and partial regression coefficient if there is more than one predictor. This value also represents standardized regression coefficients and unstandardized regression coefficients. This regression coefficient represents the number of changes that occur in Y caused by a change in the value of X. To calculate this change can be done by multiplying the observed values of the independent variable for a particular case (data) with the predictor regression coefficient.

- a: is an intercept which is a Y value when the predictor value (the independent variable) is zero (X = 0).

8.1.1.3 Linear Relationship in Linear Regression

The regression line has 3 (three) possibilities: 1) positive linear relationship, 2) negative linear relationship, and 3) no linear relationship. The picture looks like this:

1)   Positive Linear Relationship

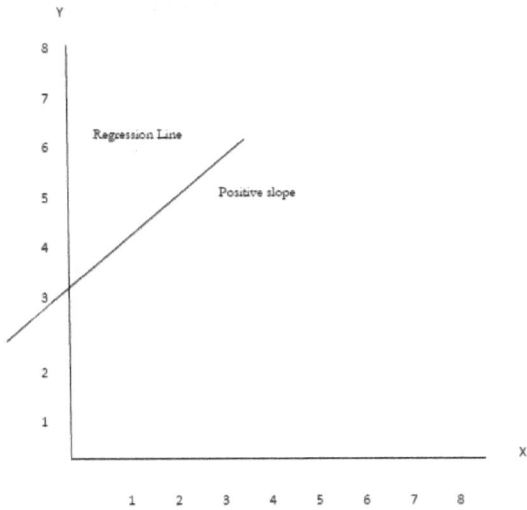

Figure 8.2 Positive Linear Relationship

2) Negative Linear Relationship

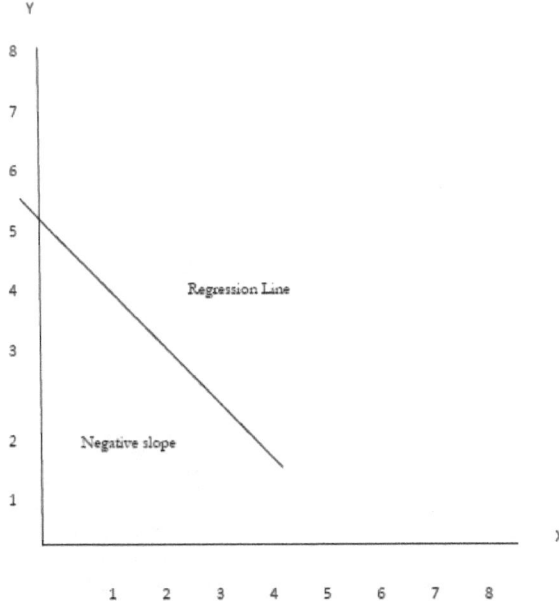

Figure 8.3 Negative Linear Relationship

3) No Linear Relations

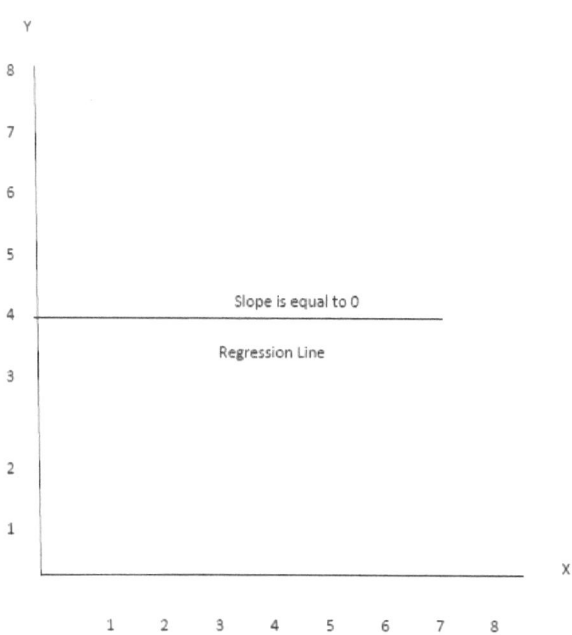

Figure 8.4 No Linear Relationship

Data Analysis and Its Interpretation: Application in Marketing Research

8.1.1.4 Variable Terms in Linear Regression

In order for us to obtain clarity in the use of the term, the following given terms that represent the meaning of independent and dependent variables. Gujarati gives the following terms:

• Dependent variable: it is also known as a predicted and / regressand / response / endegenous / outcome / response variable.

• Explanatory variable: it is also called an independent variable / predictor / regressor / exogenous / covariate / control variable.

8.1.1.5 Least Square Method in Linear Regression

Linear regression uses the Least Square method which is a way to find the line that best matches the data we have. This most suitable line can be obtained by finding any line from the lines that can be drawn, which results in the smallest number of differences between the observation data points and the line. To make it easy to understand this description in the following figure is shown when a particular line matches or fits the data; it will be a small difference between the values predicted by the line with the data resulting from the observation (research). The difference can be positive where the points of data distribution are above the line indicating where the model we created underestimates the data distribution values. Negative differences where the points of distribution of data are below the line indicating where the model that we overestimate the value of data distribution. The difference in linear regression is called residual (the difference between prediction value and observation value). For a good model the residual value should be smaller as it is depicted by the distribution of data closer to the straight line. In other words, if the dots of data both above and below the straight line are closely attached to a straight line indicating an increasingly small residual value which ultimately indicates an increasingly true model.

When we add up the positive and negative differences there will be a tendency to eliminate each other. To avoid this case, before summing up the differences, we have to squarify those differences first. These squared differences will provide a means of calculating how well a particular line matches the data. If the squared divergence is large, the line will be incompatible with data or not representing data; otherwise if the difference is small then the line will match the data or not represent the data. The sum of squared differences (SS) can be calculated for each line that matches the specified data. Goodness of Fit for each line can be compared by looking at each SS value. The Least Square method works by selecting a line having the smallest sum of square (SS) value. In other words, this method selects the best lines (matching) to represent observation data. Thus the best way is to calculate the SS value for each line until it gets the line that has the smallest SS value.

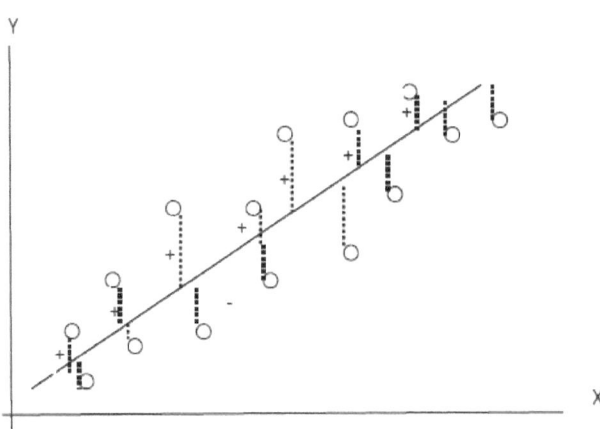

Figure 8.5 Disibution of data with lines representing a general trend

In the picture above, it shows the distribution of data with the line where the data distribution is a straight line as the general trend line (straight line from bottom left to top right showing the linear relationship between variables X and Y). The spheres describe the distribution of data, the slash represents a straight line indicating the linear relationship between X and Y while the vertical line is cut into pieces representing the difference or residual value between the line and the actual data.

8.1.1.6 Eligibility of Simple Linear Regression Model From the View Point of the R Square (R^2) Value

Is the regression model made based on our observational data correct? The answer is to look at the value of R^2. The value of R^2 is the value obtained from the sum of squares of the model divided by the sum total square. The formula is as below.

$$R^2 = \frac{SSreg}{SStotal} = 1 - \frac{SSres}{SStotal}$$

Where:

- SSreg = sum of square values of the regression model

- SStotal = SSy = total sum of square value

- SSres = sum of square residual values

The result of this R^2 calculation can be made in percentage by multiplying with 100% which then is called as Coefficient of Determination which states the percentage of variability of the dependent variable which can be explained by the regression model that we make. If the value of R^2 is rooted we will get the Pearson Correlation Coefficient (R). Thus in simple linear regression the R value gives a good estimate for the overall fit of the regression model and R^2 provides a good assessment for the fundamental measure of the relationship of the two variables we regress (Field, 2006). In simple language we can say that Pearson's correlation coefficient (R) can be used as an indicator of true or false simple linear regression model as a whole while R^2 values reflect the truth of the fundamental relationship between the independent variables and the dependent variables we regress. The conclusion is that if the value of R is higher close to 1 (the maximum value of R is 1) then the model is made more correct and the R^2 value is higher then the relationship between the independent variable and dependent variable approaches the perfect linear.

However, we cannot make the feasibility conclusions of the model merely using the values of R and R^2. Since the main purpose of creating linear regression model is not only to look for high values of R and R^2 but there is a more important goal of predicting significant regression coefficient (β) parameters. If we only look at the value of R^2 then we see the truth of the regression model that we make only from the point of view of the linearity of the two variable relations that we regress.

As Gujarati (2009) argues, he says it needs to be straightened out that the main purpose of linear regression analysis procedure is not merely to obtain the highest possible R^2 value but more to obtain the estimation of regression coefficients in the actual population through the sample we examine which ultimately that value can be used to do inference statistically. If when we do research the value of R^2 is high, it is one of the research guidelines is correct and good; even with a non-high value of R^2 does not mean that the results of the research are wrong or bad. Therefore, we should focus more on the theoretical relevance of the relationship between the independent and the dependent variables and the statistical significance of those relationships. By using a common language, we should focus more on obtaining significant research results that are far more important than obtaining a high R^2 value solely. This is also confirmed by Goldberger (1998) in his book Introductory Econometrics as quoted by Gujarati (2009) which says: "Even so high R^2 value is not proof that the model is true; otherwise low R^2 does not mean the model is wrong. In fact the most important thing in linear regression is the focus on the parameters in the population rather than the model fit in the sample."

8.1.1.7 Adjusted R Square ($\bar{R}^2$) Value in Simple Linear Regression

Adjusted R square is the R^2 values adjusted by the number of independent variables or predictors included in the regression equation and sample size. The assumption is if the independent variable is added this value tends to rise. This value is often used as the value of the goodness of fit of the model where if the value is higher (close to 1), the model is more correct / accurate. This value is generally smaller than its R square value although it can sometimes be the same (the adjusted R square ($\bar{R}^2$) ≤ the R square (R^2)). This value can only increase if the absolute t value of the added variable is greater than 1. When compared between the R square values and the adjusted R square values for model matching measurements, the value of the adjusted R square will be better. To produce a valid comparison it takes the same dependent variable. The formula for this value is:

$$\bar{R}^2 = 1 - \frac{SSres/(n-k)}{SStotal/(n-1)} = 1 - (1 - R^2)\frac{n-1}{n-k}$$

Where:
- SSres = sum of square residual value
- SStotal = SSy = total sum of square value
- n = number of sample
- k = number of variables
- R^2: R square value

8.1.1.8 The Value of F

When viewed from the formula below, then to obtain a high F value, then we need a small residual value and the amount of large data and the number of independent variables (predictors) should not many. Residual value will be small if the linear relationship between independent and dependent variables. The relationship between independent and dependent variables will be more linear if the observed data we get in the field is normally distributed and in quantities that meet the requirements, for example with a confidence level of 95% data will be between 300 - 400 and with a confidence level of 90% of data will be between 98 - 100

The value of F is directly proportional to the value of R^2, so if the value of R^2 is greater, then the value of F is also greater. Conversely, if the value of R^2 is smaller, then the value of F is also getting smaller. Therefore we can conclude that hypothesis testing with F value not only serves as a test of the overall significance of the estimated regression, but also for testing the significance of R^2 value. Thus to obtain the value of F in addition to using the formula above, we can get it by the formula:

$$F = \frac{R^2/(k-1)}{(1-R^2)/(n-k)}$$

8.1.1.9 Regression Coefficient Value (β)

Regression coefficient given β notation is a statistical value in the principal linear regression that represents the gradient or slope on the regression line. This value expresses a change in the dependent variable caused by one unit change in the independent variable (predictor). Regression coefficient is used to measure the strength of linear relationship between independent and dependent variables which then we interpret as the influence of the independent variable on the dependent variable. This regression coefficient contains 2 (two)regression coefficients, namely: standardized regression coefficient with β notation and unstandardized regression coefficient with b notation. In simple linear regression, we will use the value of the unstandardized regression coefficient (b) having the following formula.

$$b = \frac{n \sum XY - (\sum X)(\sum Y)}{n \sum X^2 - (\sum X)^2}$$

Standardized regression coefficient (β): statistical value in linear regression that describes the strength and direction of linear association between the dependent variable (criterion) and the independent variable (predictor). This value is called standardized because of its value range between -1 to 1 (Cramer & Howitt, 2006). If the value is closer to 1 then it shows the relationship between the two variables getting stronger by ignoring whether positive or negative. Thus the independent variable will be stronger to be used to predict the dependent variable. Since predictors are standardized it is possible to compare the relative strength of their relationship or weight to the dependent variable. If there is no sign (positive or negative), then the relationship between the two positive variables is interpreted. The positive correlation is significant if the value on the predictor is high, then the value on the dependent variable is also high. Conversely, if there is a negative sign, then the relationship between these two variables is negative. Negative relation has meaning if if the value on predictor is high, then the value on dependent variable is low. The coefficient with a value of 0.5 has the meaning that for every standard deviation increase in the predictor value, the standard deviation on the dependent variable rises by 0.5.

The standard regression coefficients can be converted to non-standard regression coefficients by multiplying the standard regression coefficients with standard deviation (SD) of dependent variables and dividing them by the standard deviation of the predictors. If it is formulated it will be:

Data Analysis and Its Interpretation: Application in Marketing Research

$$\text{Unstandardized regression coefficient} = \text{Standardized regression coefficient} \times \frac{\text{SD of dependent variable}}{\text{SD of independent variable}}$$

The standard regression coefficient (β) is not used in simple linear regression except in certain cases when we standardize values into Z because the variables we examine have different units of measurement from each other. This coefficient is used in the Path Analysis and Partial Least Square Structural Equation Modeling procedures.

Unstandaridzed regression coefficient (b)): an unstandardized regression coefficient has random and unlimited values opposed to the standardized coefficient with values ≤ 1. Certain predictors measured in units with larger values will have the possibility has an unstandardized partial regression coefficient that is larger than a particular predictor measured in a smaller value unit. Therefore, we will have difficulty comparing the relative weights of these predictors as they are not measured using the same scale or size. The unstandardized regression coefficient can be converted into the standardized coefficient by multiplying the unstandardized regression coefficient with standard deviation of the independent variable divided by the standard deviation of the dependent variable.

$$\text{Standardized regression coefficient} = \text{Unstandardized regression coefficient} \times \frac{\text{Standard deviation of the independent variable}}{\text{Standard deviation of the dependent variable}}$$

In simple linear regression, the regression coefficient we use is an unstandardized regression coefficient (b).

8.1.1.10 The t value

The value of t is obtained in the regression coefficient output section which serves to be used as partial or individual hypothesis testing when we use simple and multiple linear regression procedure where we use more than one independents variable or predictors. At the moment we formulate hypothesis and do hypothesis testing with the value of t then the statement of the null hypothesis is the value of β is equal to 0. Thus if the regression coefficient is significant which means the value is not equal to 0, then the independent variable contributes significantly to the change in the value of the dependent variable.

The criteria of hypothesis testing is done by comparing the t value (t_o) with t table (t critical value) using the provisions, namely: if the value of t_o > t table with a certain level of significance (α), for example of 0.05 then H0 is rejected and H1 accepted; otherwise if the value of t_o < t table then H0 is accepted and H1 is rejected. To obtain the value of t used the formula as below

$$to = \frac{b_k}{Sb_k}$$

with $Sb = \frac{Se}{\sqrt{\Sigma x^2}}$

$$Se = \sqrt{\frac{\Sigma Y^2 - a\Sigma Y - b\Sigma XY}{n-2}}$$

$$b = \frac{n\Sigma XY - (\Sigma X)(\Sigma Y)}{n\Sigma X^2 - (\Sigma X)^2}$$

8.1.1.11 Probability Value (p value) at ANOVA Output

This probability value, which is in IBM SPSS called as significance (sig), can be used as an alternative to simultaneous hypothesis testing with F test with the following conditions: if the probability value of the calculation result < alpha value (α), for example 0.05; then H0 is rejected and H1 accepted; otherwise if the probability value of the calculated result > alpha value (α), for example 0.05; then H0 is accepted and H1 is rejected.

8.1.1.12 Probability Value (p value) Regression Coefficient Output

This probability value, which is in IBM SPSS called as significance (sig), can be used as an alternative to partial hypothesis testing with t

Data Analysis and Its Interpretation: Application in Marketing Research

test with the following conditions: if the probability value of the calculation result < alpha value (α), eg 0.05; then H0 is rejected and H1 accepted; otherwise if the probability value of the calculated result > alpha value (α), for example 0.05; then H0 is accepted and H1 is rejected

8.1.1.13 Basic Terms

Terms uses linear regression in IBM SPSS based on the following:

a. Regression model is said to be feasible if the value of probability value ir significance on ANOVA is < 0.05

b. The predictor used as an independent variable should be feasible. This feasibility is known if the Standard Error of Estimate < Standard Deviation

c. The regression coefficient must be significant. Testing is done by t test. The regression coefficient is significant if t_o > t table (critical value). In IBM SPSS can be replaced by using the probability or significance value (sig) with the following conditions:

- If sig < 0.05; the regression coefficient is significant

- If sig > 0.05; the regression coefficient is not significant

d. There should be no multicollinearity, meaning there is no very high or too low correlation between the independent variables. This requirement applies only to multiple linear regression with more than one independent variables. Multicollinearity occurs when the correlation coefficient between the independent variables is ≥ 0.9

e. There must no autocorrelation. There is no autocorrelation when the Durbin Watson value: $-2 \leq DW \leq 2$

f. The goodness of fit of the regression model can be explained by using the r^2 value where the greater the value the better the model. If the value close to 1 then the regression model is better. The value of r^2 has characteristics such as: 1) always positive, 2) Maximum value is 1. If value of r^2 is equal to 1 the model will have perfect fit. That is, all variations in the dependent variable (variable of Y) can be explained by the regression model. Conversely, if r_2 is equal to 0, then there is no linear relationship between the independent variable (X) and dependent variable (Y).

g. There is a linear relationship between the independent variable (X) and the dependent variable (Y)

h. Data must be normally distributed

i. It uses interval or ratio data

j. There is a dependency relationship, meaning one variable can identified as a dependent variable that depends on other variable, namely independent variable.

8.1.1.14 Assumptions

Assumptions that underlie the linear regression are:

- The independent variable or predictor should not be correlated with the error or disturbance term if this happens then Least Square method (LS) used to estimate unknown parameters, i.e. regression coefficients will be biased and inconsistent

- Independent and dependent variables should be quantitative and metric scale. It can be interval or ratio scale

- There is no multicollinearity. This means that there is no perfect linear relationship between the predictors or the independent variables. The perfect linear relationship is reflected in the very high correlation between independent variables according to Hair (2010) of $\geq$ 0.9. Other statistical values used to test multicollinearity include: the value of variance inflation factor (VIF) with the provision if the value of VIF > 5, then the multicollinearity occurs; value condition index with condition if condition value index > 5, then happened multicollinearity. This assumption applies only to multiple linear regressions when multiple variables are more than one.

- The independent variable used as the predictor should not be correlated with external variables that are not included in the regression model we create.

- No heteroscedasticity should occur. To understand the meaning of heteroscedasticity is necessary to understand in advance the notion of homoscedasticity. Homoscedasticity is the description of data where the error terms appear to be constant beyond the reach of certain independent variable values. The assumption of the equality of the population error variance of ε (where ε is estimated from the sample value of e) is critical if applied to the correct linear regression. When the error limit has an increasingly large variance, then the data is called heteroscedasticity. In other words, homoscedasticity is the assumption that the dependent variable shows the same level of variance for all the independent variables. If the spread of the variance value on all independent variables is not the same then the relationship is said to be heteroscedasticity. To test homoscedasticity (the occurrence of variance equality in all independent variables) Levene testing is used on non-metric variables. Variance equality occurs when the value of significance (sig) in Levene test is > 0.05. Levene testing can be explained by making the null hypothesis (H0) which states "variance on all independent variables are equal" and alternative hypothesis (H1) which states "variance on all independent variables are not equal". The provision of hypothesis testing is based on the value of significance: if sig value > 0.05 H0 is accepted; if sig value < 0,05 H0 is rejected. If the metric-scale variables we can use the test Box's M. Terms of testing the same as the test using Levene test. To find out whether the occurrence of heteroscedasticity in the above test can be known from the significance value. If the significance value (sig) <0.05, then in the heteroscedasticity in the model occurs. If using graphs, heteroscedasticity occurs in the regression model if the points in the scaterplot form certain patterns or assemble on one side or near the value 0 on the Y axis on the resulting curve when we draw the curve using SPSS. If the data points are not spontaneously distributed, there is no heteroscedasticity.

- No autocorrelation occurs. Autocorrelation occurs in independent variables that interfere with the relationship of independent variables with dependent variables. Autocorrelation occurs when two observations or values correlate in one variable. For autocorrelation testing we use values from Durbin - Watson (DW). The range of DW ranges from 0 - 4. There is no autocorrelation if the value of DW: $-2 \leq DW \leq 2$ (Anderson, 2001: 733)

- Errors that are normally distributed. It is assumed that the residuals in the model are random, the variables are normally distributed with an average of 0.

- Linearity. It means that the mean value of the dependent variable for each increase of the independent variable lies in a straight line. In other words the regression model that we make is linear.

8.1.1.15 Using Linear Regression in IBM SPSS

To use the linear regression procedure in IBM SPSS we need the following steps.

First: Enable SPSS, go to Data Editor and select sub-menu in bottom left of Variable View. After that create a design variable for both variables that we use in research by filling information in accordance with Default Variable View on the columns available as in the example below.

Name	Type	Width	Decimal	Label	Values	Missing	Column	Align	Measure	Role
Month	String	8	0	Month	None	None	8	R	Nominal	Input
Sales force	Numeric	8	2	Number of sales force	None	None	8	R	Scale	Input
Sale	Numeric	8	2	Sales	None	None	8	R	Scale	Target

When finished entering the requested information then click the Data View command located next to the Variable View command

Second: Enter data in position of Data View

Go to the Data View command, where the command is read as follows:

- Columns are read as variables and rows are read as cases

Fill the data into 1 on the first line up to the 12th data on the twelfth line

Data Analysis and Its Interpretation: Application in Marketing Research

from top left to bottom as shown below:

	Month	Sales force	Sale
1	January	20	200
.			
.			
12	December	14	60

Third: Conduct analysis

After completion of entering all the data, we will do the data analysis using the steps as follows.

- Click Analyze > Regression: select Linear

- Move the dependent variable of the sale to the Dependent Variable column

- Move the independent variable of the sales force to the Independent variable column

- Enter the month variable into the Case Labels column

- Fill in the Method field with the Enter command

- Click on Option: In the option Stepping Method Criteria enter the number 0.05 on the Entry column (meaning that the confidence interval that we use is 95%)

- Check Include constant in equation

- In the Missing Values option check Exclude cases listwise > Press Continue

- Select Statistics: In the Regression Coefficient option select Estimate (to calculate the regression coefficient value of b), Fit

Model (to calculate the value of F, R, R^2, and Adjusted R^2) and Descriptive (to calculate the mean value, standard deviation and correlation). In the Residual option (to exclude the difference between the observed value and the predicted value), select Case wise Diagnostics and check All Cases (to generate predicted value according to the amount of data) > Press Continue

- Click Plots to create Graphics

- Fill in column Y with SDRESID (studentized residual) option and X column with ZPRED (default variable prediction value depends based on model), then press Next

- Fill in column Y with ZPRED and X column with DEPENDNT (output variable)

- In the Standardized Residual Plots option (to generate the Standard Residual Curve), check the Normal Probability Plot (to make the Normal Probability Curve) > Press Continue

- Click OK

* Note: In IBM SPSS there are two main methods, namely Enter method and Stepwise method consisting of Stepwise, Forward and Backward used for Multiple Linear Regression. The Enter method has the meaning of all variables calculated based on IBM SPSS criteria (default). In simple linear regression we use the Enter method. Stepwise method has the understanding of the relationship between independent variables with dependent variables based on the extent to which these variables are highly correlated with dependent variables. The Forward method in IBM SPSS is the same as Stepwise where the difference in the Forward method will be the removal of the independent variables that have less significant impact on the dependent variable; while the Backward method prioritizes calculations based on the order of significance of the independent variables used as predictors. The independent variables that contribute the most significant will take precedence and then followed by significant and signi fi cant variables. Selection of this method applies only when we use more than one independent variable in multiple linear regression.

Data Analysis and Its Interpretation: Application in Marketing Research

The results of the analysis will be listed in the following sections. We will proceed to step 4, which is to make interpretation of the result of analysis by using linear regression procedure.

Fourth: Make interpretation of the analysis result.

Part I: Descriptive Statistics

Descriptive Statistics

	Mean	Std. Deviation	N
Sale	147.83	71.978	12
Number of salesforce	20.17	8.277	12

This section is used to interpret the average of sale and the amount of sales force.

- Average of sale is 147.83 and average of sales force is 20.17 for twelve months.

- Standard Deviation of the sale variable is equal to 71.978 while for sales force is equal to 8.277. The standard deviation value is a standard deviation value indicating the data dispersion; thus the value can be interpreted if the value is getting smaller (close to 0) then this value indicates data that is homogeneous. If the unit of observation data values in large numbers, for example millions then at least the standard deviation value is smaller than the average value.

- Number of cases (N) of 12

Part II: Correlation Between Variables of Sales Force and Sale

In using linear regression the reader must understand that IBM SPSS will produce an output the correlation value between the two variables under study. Although the result of correlation will not be used to answer the problem formulation above; we will discuss the value of correlation and its interpretation. A correlation value in simple linear regression is the value of Pearson correlation coefficient obtained by calculating correlation coefficient value by using Pearson Correlation formula as follows:

$$r = \frac{n \sum XY - (\sum X)(\sum Y)}{\sqrt{(n \sum X^2 - (\sum X)^2) - (n \sum Y^2 - (\sum Y)^2)}}$$

In addition we can also calculate it by rooting the value of R^2; therefore the value of correlation coefficient (R or r) is useful also to see the feasibility of simple linear regression model that we make. Provision if the value of R is higher then the regression model more feasible. This is easy to understand because if the value of R is high, then the relationship of the two variables is getting linear (forming a straight line).

The main uses of the Pearson correlation coefficient are:

• First: measure the relationship strength of two variables

• Second: see the significance of the relationship of two variables

• Third: look at the relationship direction of two variables. There are two types of relationship directions in the correlation, ie direct or positive relationships (unidirectional) or negative relationships.

Correlations

		Sale	Number of salesforce
Pearson Correlation	Sale	1.000	.716
	Number of salesforce	.716	1.000
Sig. (1-tailed)	Sale		.004
	Number of salesforce	.004	
N	Sale	12	12
	Number of salesforce	12	12

The above output has the following meaning.

- The value of correlation between a sales force and sale variables is 0.716. This means that the relationship between these two variables is strong because the value is greater than 0.5. The correlation ranges between 0 - 1 and the correlation coefficient can be positive (+) or negative (-); so that the range can be formulated to be between -1 upto 1. If the correlation coefficient is found positive then the relationship of the two variables has positive correlation; otherwise if the correlation coefficient is found negative then the relationship of the two variables has anegative correlation.

When the correlation value between two variables is is closer to 1, the correlation between the two variables is said to be stronger and stronger; on the contrary if the correlation is getting closer to 0, then the correlation between the two variables is getting weaker. The following criteria can be used to identify the magnitude of the correlation value.

- 0: There is no correlation between two variables

- $> 0 - 0.25$: The correlation is very weak

- $> 0.25 - 0.5$: The correlation is moderate

- $> 0.5 - 0.75$: The correlation is strong

- $> 0.75 - 0.99$: The correlation is very strong

- 1: Perfect correlation

The correlation between sales force and sales variables is significant when viewed from the significance value (sig) of 0.004 smaller than 0.05. If the number of significance < 0.05, it means that there is a significant correlation between the two variables. In addition if the correlation coefficient output table is indicated by two stars (**); then the significance (sig) of the comparator is not 0.05 but 0.01. The resulting positive correlation coefficient (0.716) has the meaning that the correlation between the two variables is unidirectional. This means that if the sales force value increases, then sale valu will increase as well.

Part III Variable Entered

Variables Entered/Removed[a]

Model	Variables Entered	Variables Removed	Method
1	Number of salesforce[b]	.	Enter

a. Dependent Variable: Sale
b. All requested variables entered.

This section shows the methods in entering variables

In this section we include all the variables that will be analyzed and no variables that we remove because we use the method "Enter". Other methods provided in the Method command are a) stepwise, b) forward, and backward

Part IV Model Summary (Coefficient of Determination)

Model Summary[b]

Model	R	R Square	Adjusted R Square	Std. Error of the Estimate	Durbin-Watson
1	.716[a]	.513	.464	52.702	2.118

a. Predictors: (Constant), Number of salesforce
b. Dependent Variable: Sale

The above output shows the amount of R^2 obtained by using the following formula:

$$R^2 = \frac{SSreg}{SStotal} = 1 - \frac{SSres}{SStotal}$$

To facilitate the interpretation we make the value into percent form which we then call as Coefficient Determination. The function is to know the amount of proportion of variation of the dependent variable of sale which can be explained by using the independent variable of sales force. Coefficient of determination is calculated by multiplying the value of R^2 by 100% (R^2 x 100%)

• The R Square value is 0.513 (51.3%). This figure has an understanding that as much as 51.3% proportion of the sales that occur can be explained by using the independent variable of sales force. In common language we can say that the independent variable of the sales force affects the dependent variable of the sale. The remainder as much as 48.7% or (100%

- 51.3%) should be explained by other factors beyond this regression model under study.

As we know that the magnitude of R square ranges from 0 -1 which means the smaller the magnitude of R square, then the variation in the dependent variable can be explained less by using the independent variables that we choose. Conversely, if R square is close to 1, then the dependent variable can be explained more by using the independent variable under study.

The amount of Error of the Estimate Standard (SEE) is 52.702 (for the sale variable). If the number is compared to the Standard number Deviation (STD) which is equal to 71.978, then this SEE number is smaller that STD. Thus the independent variable of sales force can used as predictor in determining the sale is correct. The rule is that the predictor for the dependent variable is correct if the SEE value is less than the standard deviation (SEE <STD).

Section V Anova: Feasibility Test of the Regression Model

This output is used to test the feasibility of the regression model under study.

ANOVA[a]

Model		Sum of Squares	df	Mean Square	F	Sig.
1	Regression	29214.496	1	29214.496	10.518	.009[b]
	Residual	27775.170	10	2777.517		
	Total	56989.667	11			

a. Dependent Variable: Sale
b. Predictors: (Constant), Number of salesforce

This section shows the magnitude of the F value and the probability or significance value (Sig) in the Anova calculation which will be used for the feasibility test of the regression model with the following provisions:

If using the F value then the value of F observation (Fo) must be greater than F table (Fα)

If using the significance value, then a good model should have a significance value < 0.05.

The output above generates:

The output of ANOVA produces a F number of 10.518 with a significance level (probability number) of 0.009. Since the significance level value is 0.009 < 0.05, thus the regression model is feasible to be used in predicting the sale. To be used as a regression model that can be used in predicting the dependent variable, then the significance value (sig) should be less than 0.05 (sig < 0.05).

To perform the feasibility test of the regression model we created with the data we have, the steps are as follows:

Using the value of F

To perform hypothesis testing with the value of F, the steps are as follows

First: Make a hypothesis

H0: $\beta = 0$ (the sales force variable does not significantly affect the sale variable

H1: $\beta \neq 0$ (the sales force variable significantly affects the sale variable

Second: Calculate the value of F observation (Fo) with the following formula

$$F = \frac{SSreg/(k-1)}{SSres/(n-k)}$$

From the calculation of IBM SPSS it is obtained the Fo value as much as 10.518.

Third: Use the following criteria:

If the value of Fo > F table (Fα); then H0 is rejected and H1 accepted

If the value of Fo < F table (Fα); then H0 is accepted and H1 is rejected

Fourth: Calculate the value of F table obtained from Table of F value with the following conditions:

Use an alpha value of 0.05 (α) and Degree of Freedom (DF) for numerator = number of variables - 1 or (k-1) and denominator = number of data - number of variables or (n - k). Thus the value of F table for the numerator is: 2 - 1 = 1 and the denominator is: 12 - 2 = 10; then obtained value of F from table equal to 4.96.

Fifth: Make a decision

Obtained from the calculation of the F value is 10.518 > F table value of 4.96; then H0 is rejected and H1 accepted. This means that the variable of sales force has significant effect on the sale variable.

Hypothesis testing with F value can also be done by using curve as follows:

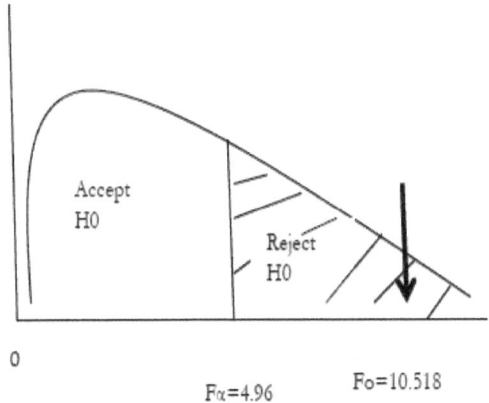

The curve image shows that the Fo value is in the rejected H0 area so the H1 is accepted. This means that the variable of sales force has a significant effect on the sales variable.

Use the value of significance

First: Determining the hypothesis

H0: $\beta = 0$ (the sales force variable does not significantly affect the sale variable)

H1: $\beta \neq 0$ (the sales force variable significantly affects the sale variable)

Second: Using the following decision-making criteria

- If the value of significance < 0.05; then H0 is rejected and H1 accepted

- If the value of significance > 0.05; then H0 is accepted and H1 is rejected

Third: Make a decision

From the output table above, it is known that the significance value in Sig column is 0.009 < 0.05; then H0 is rejected and H1 accepted. This means that the variable of sales force has a significant effect on the sale variable.

Both using the value of F and the value of significance indicate the same result. In practice, we can use either the F value or the significance value.

Section VI Regression Coefficients

This output is used to test the hypothesis of regression coefficient significance. Regression coefficients represent slope or gradient in the regression line. This value represents the change in the dependent variable as a result of the one unit change in the independent variable or its predictor. In other words, the amount of change in the predictor will change (can be positively or negatively) the value of the dependent variable. The regression coefficients to be used in simple linear regression are unstandardized coeficients or "b" (in simple linear regression we can use "b" notation only because the number of independent variable is only 1) which in IBM SPSS is given the symbol of "B" located in the unstandardized column coeficients.

Another regression coefficient is a constant value (Constant) having "a" or "b0" notation which in IBM SPSS is given a notation of "B". This value has the meaning of the dependent variable value (Y) when the independent variable value is 0 (when X = 0). This value can be positive and negative.

Coefficients[a]

Model		Unstandardized Coefficients		Standardized Coefficients	t	Sig.
		B	Std. Error	Beta		
1	(Constant)	22.276	41.596		.536	.604
	Number of salesforce	6.226	1.920	.716	3.243	.009

a. Dependent Variable: Sale

This section describes the regression equation to find the constant value, and test the hypothesis of regression coefficient significance:

The regression equation is:

$Y = a + b x$

Where:

- Y = sale

- X = sales force

- a = constant number of Unstandardized Coefficient which in this research is equal to 22.276. This number is a constant number which means: if there is no additional sales force (X=0), then sale will be as much as 22.276 (22)

- b = the regression coefficient value is 6.226. This number means that in every addition of 1 sales force, then sales will increase by 6.226. Conversely, if this value is negative (-) then sale will decrease by that number.

Therefore, the equation becomes:

$Y = 22.276 + 6.226 X$

To perform hypothesis testing we can do by using t value (to) or significance value (sig or α) with the steps as follows:

Testing by using t value

First: Determining the hypothesis.

H0: b = 0 (the sales force variable does not significantly affect the sale variable)

H1: b ≠ 0 (the sales force variable significantly affects the sale variable)

Second: Calculate the value of t observation (t_o)

From the calculation using IBM SPSS it is obtained value of 3.243

Third: Use the following criteria:

If t value of observation (to) > t table (tα); then H0 is rejected and H1 accepted

If t value of observation (to) < t table (tα); then H0 is accepted and H1 is rejected

Fourth: Calculate the value of t table obtained from t table with the following conditions:

The value of α is 0.05 and Degree of Freedom (DF) = n -2. Because testing is done on two sides (ie b = 0 or b ≠ 0); then the value of α must be subdivided by 2 so the value becomes $\alpha / 2 = 0.05 / 2$ the result is 0.025. The value of t_α 2.28.

Fifth: Make a decision

Obtained from the calculation of t count value of 3.423 > t table value of 2.28; then H0 is rejected and H1 accepted. This means that the variable of sales force has a significant effect on the sale variable.

The hypothesis testing with t value can also be done by using curve as follows:

Data Analysis and Its Interpretation: Application in Marketing Research

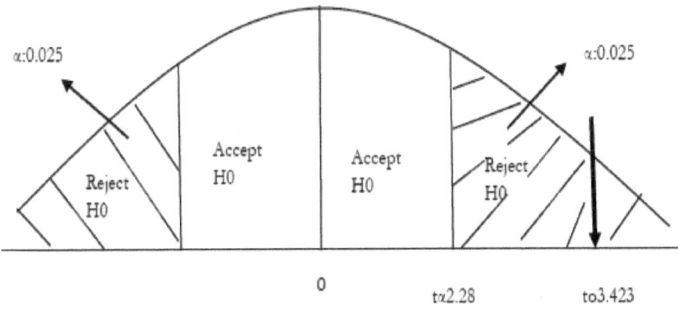

If the value t observation (t_o) is positive, then the test is done on the right side; otherwise if the value t_o is negative, then the test is done on the left side. On the curve above we do hypothesis testing on the right side because the result of t value is positive. We see that the value of t_o is in the area of H0 rejection, thus H1 is accepted. This means that the variable sales force has a significant effect on the sale variable.

Testing with significance value (sig)

As shown in the above output IBM SPSS also generates signficance value (sig) in addition to the t value that we can use as another alternative to perform hypothesis testing with the t value.

First: Determining the hypothesis

H0: the sales force variable does not significantly affect the sale variable

H1: the sales force variable significantly affects the sale variable

Second: Using decision-making criteria

If the value of significance (Sig) < 0.05; then H0 is rejected and H1 accepted

If the value of significance (Sig) > 0.05; then H0 is accepted and H1 is rejected

Third: Make a decision

From the output above, it is known that the significance value in Sig column is 0.009 <0,05; then H0 is rejected and H1 accepted. This means that the variable of sales force has a significant effect on the sale variable.

Whether using t value or significance value result of the hypothesis testing will be the same.

Note: In simple linear regression testing the model feasibility hypothesis and regression coefficient significance are the same. This will be different in multiple linear regression where in model feasibility testing using F value. The significance value on ANOVA output is referred to as simultaneous testing; while the significance value at regression coefficients on Coefficients output is referred to as partial test.

Section VII Diagnosis by Case

The following output is the output of the predicted and residual values or the difference between the observed value and the predicted value. The prediction value is the value obtained by using the linear regression equation Y = a + b x. The residual value is obtained by means of observation value minus the prediction value.

Data Analysis and Its Interpretation: Application in Marketing Research

Casewise Diagnostics[a]

Case Number	Month	Std. Residual	Sale	Predicted Value	Residual
1	January	-.172	200	209.06	-9.056
2	February	-.720	140	177.93	-37.926
3	March	.272	130	115.67	14.334
4	April	.063	175	171.70	3.300
5	May	.469	240	215.28	24.718
6	June	-1.335	120	190.38	-70.378
7	July	1.432	160	84.54	75.464
8	August	.222	90	78.31	11.690
9	September	1.844	300	202.83	97.170
10	October	-.152	89	96.99	-7.988
11	November	-.985	70	121.89	-51.892
12	December	-.938	60	109.44	-49.440

a. Dependent Variable: Sale

The above output shows the result of prediction of regression equation. The discussion starts from Case Number 1, for January, then the regression equation is:

- $Y = a + b x$

$Y = 22.276 + 6.226 \times 30$

- For January the amount of sales force (variable X) based on observation data is equal to 30, then the equation becomes as follows: $Y = 22.276 + (6.226 \times 30)$. The result if calculated manually is 209.056 This figure is equal to the predicted value in January as calculated by IBMSPSS, i.e. 209.06. For other case calculations in the next months we can see the output in the Predicted Value column

- The residual column provides an explanation of the difference between the actual sale and the predicted sale or $200 - 209.06 = -9.06$. If the difference gets smaller then predictions are more accurate; on the contrary if the difference is greater then predictions are increasingly inaccurate.

- Standardized residual denotes residual divided by Standard Error of the Estimate (SEE). For the first case: $-9.06 / 52.702 = -0.172$. The SEE value as much as 52.702 comes from the interpretation of model summary.

- The amount of residual and standardized residual give meaning to the regression equation that will be used to predict the dependent variable value. The smaller the residual and standardized residual values, the better the regression model to use in predicting.

- The tendency of sale is to increase and decrease when viewed at the Residual column as the difference between the observation value and the predicted value.

Section VIII Residual Statistics

Residuals Statistics^a

	Minimum	Maximum	Mean	Std. Deviation	N
Predicted Value	78.31	215.28	147.83	51.535	12
Std. Predicted Value	-1.349	1.309	.000	1.000	12
Standard Error of Predicted Value	16.900	26.287	21.259	3.459	12
Adjusted Predicted Value	63.19	211.49	145.93	52.962	12
Residual	-70.378	97.170	.000	50.250	12
Std. Residual	-1.335	1.844	.000	.953	12
Stud. Residual	-1.444	2.045	.016	1.048	12
Deleted Residual	-82.341	119.501	1.905	60.792	12
Stud. Deleted Residual	-1.540	2.543	.062	1.173	12
Mahal. Distance	.214	1.820	.917	.590	12
Cook's Distance	.000	.480	.107	.159	12
Centered Leverage Value	.019	.165	.083	.054	12

a. Dependent Variable: Sale

This section provides an explanation of the minimum value of the predicted sale, namely: 78.31; the maximum value is 215.28; average of the sale is predicted to be 147.83.

Section IX Chart 1: Normal Probability Plot

Data Analysis and Its Interpretation: Application in Marketing Research

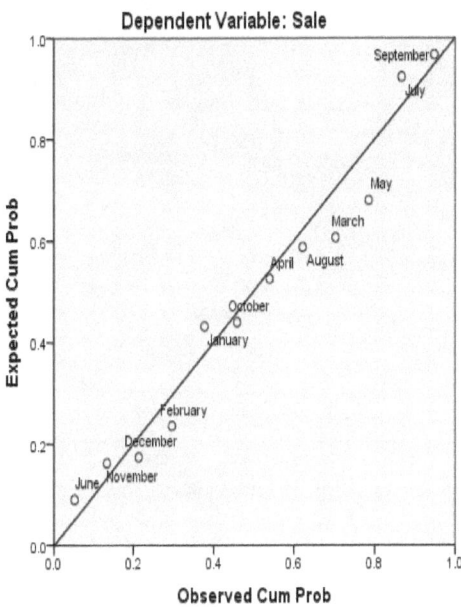

The graph above shows the fulfillment of the normality requirements of data distribution, i.e. if the residual is from a normal distribution, then the data distribution values will be in the area around the straight line as discussed above section that the use of simple linear regression is to estimate the value of dependent variable based on the straight-line equation. From the calculation results we see the graph above shows that the distribution of data is in position around the straight line forming a slash from the lower left to the top right; therefore the normality requirements have been met.

Section X Graph 2: Regression Fit Model

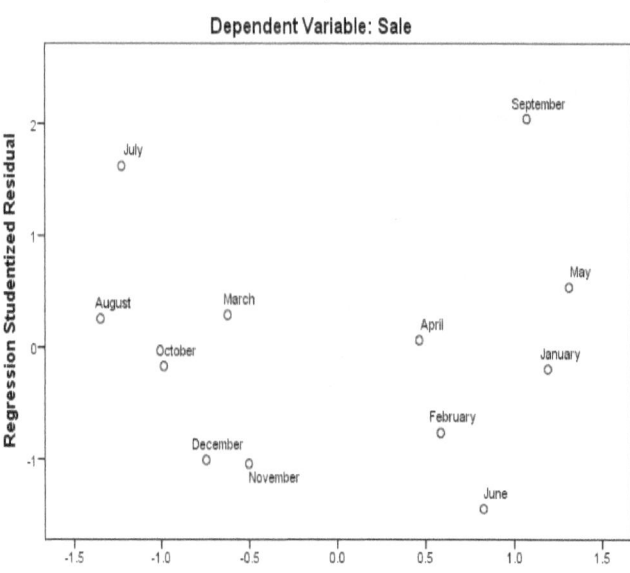

The graph above gives an explanation of the relationship between predicted value (the sale) and Studentised Delete Residual respectively. The description is as follows: Regression model is feasible used to predict if scattered data is scattered around the number 0 (zero) on the Y axis and does not form a particular pattern or trend. If we see the data distribution above is in the zero point area of Y axis, then this regression model is feasible to be used to predict the sales. From the picture above only three data are not located around the point 0, i.e. cases in August and September that are located in the extreme at the top right.

Section XI Chart 3: Model Fit Requirements for Each Data

Data Analysis and Its Interpretation: Application in Marketing Research

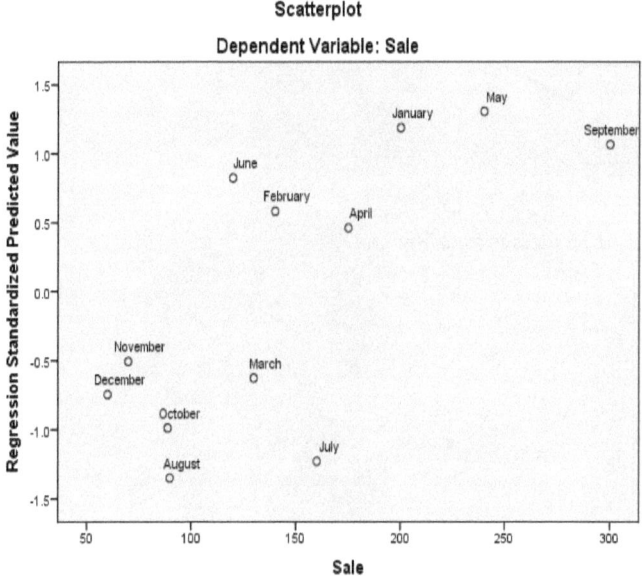

The graph above shows the relationship between the sale variable value and its prediction value. The eligible model is the distribution starting from the lower left and then straight towards the right and up. If it is seen the distribution of data above, it can be concluded that the distribution of data already follows the alignment model requirements of each data. In conclusion, this regression model is feasible to be used in predicting the sales.

8.2 The Second Case

In this case we will add another independent variable, namely number of retailers. Thus the variable relationship becomes as follows:

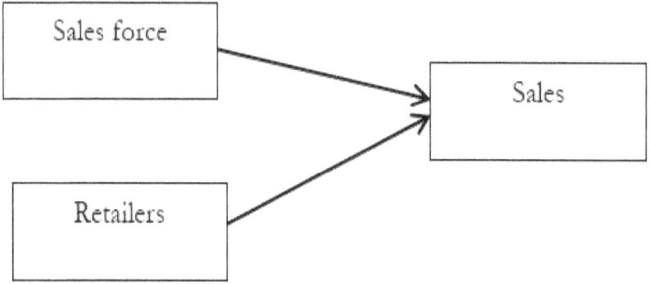

The data of number of retailers is as follows

Retailers
8
7
5
7
8
7
3
3
8
4
5
5

The formulation of the problem as follows:

- How much is the effect of sales force and number of retailers on sale partially and simultaneously?

8.2.1 Solution

If seen from the model relationship of the above variables, then we can identify as follows:

- The variables of sales force and are retailers are independent variables.

- The variable of sale is a dependent variable.

Data Analysis and Its Interpretation: Application in Marketing Research

- The variable of month is a case label

To solve this problem we will use a multiple linear regression procedure that match the problem in the study.

8.2.1.1 Definition of Multiple Linear Regression

Multiple linear regression has the same meaning as the simple linear regression. The difference lies only on the number of independent variables or predictors. In multiple linear regression, there are more than one independent variables.

8.2.1.2 Methods Available in Multiple Linear Regression

In IBM SPSS there are two main methods, namely Enter method and Stepwise method consisting of Stepwise, Forward and Backward) used for Multiple Linear Regression. The Enter method has the meaning of all variables calculated based on IBM SPSS criteria (default). In simple linear regression we use the Enter method. Stepwise method has the understanding of the relationship between the independent variable with the dependent variable based on the extent to which these variable are highly correlated with the dependent variable. The Forward method in IBM SPSS is the same as Stepwise, the difference is in the Forward method there will be removal of the independent variables that have less significant impact on the dependent variable; while the Backward method prioritizes calculations based on the order of significance of the independent variables used as predictors. The independent variables that contribute the most significant will take precedence and then followed by less significant variables. Selection of this method applies only when we use more than one independent variables in multiple linear regression. The choice of the safest method is the Enter method. The stepwise method will produce the independent variables available in the reduced model because it is exposed to the removal process by the IBM SPSS program, i.e. the independent variables that are less significant impact on the dependent variable in the model.

8.2.1.3 Multiple Linear Regression Procedures in IBM SPSS

To use multiple linear regression procedures in IBM SPSS, the steps are:

First: Activate the IBM SPSS

After the IBM SPSS is active, then go to Data Editor and select sub-menu in bottom left of Variable View. After that create a additional variable design for the number of retailers variable.

Second: Enter data in position of Data View

Third: Conduct analysis

How to do analysis in multiple linear regression is as follows

- Click Analyze > Regression > Linear

- Move the variables of sale to the Dependent column

- Move the variables of sales force and number of retailers to the Independent column

- Enter the month variable into the Case Labels column

- Fill in the Method field with the Enter command

- Click Option: In the option Stepping Method Criteria enter the number 0.05 on the Entry column

- Check Include constant in equation in the Missing Values option check Exclude cases listwise > Press Continue

- Select Statistics: In the Regression Coefficient option select Estimate, Model Fit and Descriptive > Press Continue

- Click Plots to create a Graph, select produce all partial plots > Click Continue

- Click OK

Data Analysis and Its Interpretation: Application in Marketing Research

Fourth: Interpret the results of calculations

Part I Descriptive Statistics

Descriptive Statistics

	Mean	Std. Deviation	N
Sale	147.83	71.978	12
Number of salesforce	20.17	8.277	12
Number of retailers	5.83	1.899	12

This section is used to interpret the average sale, number of sales force and number of retailers variables.

• The average of sale is 147.83 and the number of sales force is 20.17 and the number of retailers is 5.83.

• Standard Deviation of the of sale is 717.978 and the number of sales force is 8.277 and the number of retailers is 1.899

Part II Correlation

Correlations

		Sale	Number of salesforce	Number of retailers
Pearson Correlation	Sale	1.000	.716	.680
	Number of salesforce	.716	1.000	.985
	Number of retailers	.680	.985	1.000
Sig. (1-tailed)	Sale	.	.004	.007
	Number of salesforce	.004		.000
	Number of retailers	.007	.000	.
N	Sale	12	12	12
	Number of salesforce	12	12	12
	Number of retailers	12	12	12

This section is to determine whether or not there is correlation between the variables of sales force and number of retailers with the sale variable The correlation between the variables of sales force and the sale is 0.716. This means that the relationship between the two variables is strong. A positive correlation indicates that the correlation between the two variables is unidirectional. While the correlation between the number of retailers and the sale is 0.680. This means that the relationship between the two variables is strong as well. A positive correlation indicates that the correlation between the two variables is unidirectional.

The correlation is significant because the significance (sig) values as much as 0.004 and 0.007 are less than 0.05.

Part III Variable Included

Variables Entered/Removed[a]

Model	Variables Entered	Variables Removed	Method
1	Number of retailers, Number of salesforce[b]	.	Enter

a. Dependent Variable: Sale
b. All requested variables entered.

This section shows the methods in entering variables. In this section we see the variables of "number of retailers" and "number of sales force" because we use the method of "Enter" which has the meaning we enter two independent variables simultaneously.

Part IV Model Summary (Coefficient of Determination)

Model Summary[b]

Model	R	R Square	Adjusted R Square	Std. Error of the Estimate	Durbin-Watson
1	.731[a]	.534	.431	54.316	2.312

a. Predictors: (Constant), Number of retailers, Number of salesforce
b. Dependent Variable: Sale

The R Square value is 0.534 (53.4%). This figure has an understanding that as much as 53.4% proportion of the sales that occur can be explained by using the independent variables of sales force and number of retailers.

Data Analysis and Its Interpretation: Application in Marketing Research

The value of the standard error of estimate (SEE) is 54.316 (for the variable of the prediction cost). When compared with the Standard Deviation (STD) as much as 71.918, the SEE is smaller than STD. This means that the SEE value shows both the independent variables are already correct predictors.

Part V Anova

This section is used to see the feasibility of the regression model and also called a simultaneous hypothesis testing by using the value of F or the value of significance.

ANOVA[a]

Model		Sum of Squares	df	Mean Square	F	Sig.
1	Regression	30437.357	2	15218.679	5.158	.032[b]
	Residual	26552.310	9	2950.257		
	Total	56989.667	11			

a. Dependent Variable: Sale
b. Predictors: (Constant), Number of retailers, Number of salesforce

In this case we use the probability or significance (sig) value in the calculation of Anova which will be used to test the feasibility of the regression model of the simultaneous hypothesis testing with the provision of a good probability number to be used as a regression model is to be less than 0.05.

The ANOVA test yields a F value as much as 5.158 with a significance level (probability) of 0.032. Since the significance value as much as 0.032 < 0.05, thus this regression model is feasible to be used in predicting the sale value.

To perform the feasibility test of the regression model we make a simultaneous hypothesis testing with the following steps:

First: Make a hypothesis

H0: $\beta = 0$ (the sales force and number of retailers variables do not significantly affect the sale variable

H1: β ≠ 0 (the sales force and number of retailers variables significantly affects the sale variable

Second: Calculate the value of F observation (Fo) with the following formula

$$F = \frac{SSreg/(k-1)}{SSres/(n-k)}$$

From the calculation of IBM SPSS it is obtained the Fo value as much as 5.158

Third: Use the following criteria:

If the value of Fo > F table (Fα); then H0 is rejected and H1 accepted

If the value of Fo < F table (Fα); then H0 is accepted and H1 is rejected

Fourth: Calculate the value of F table obtained from Table of F value with the following conditions:

Use an alpha value of 0.05 (α) and Degree of Freedom (DF) for numerator = number of variables - 1 or (k-1) and denominator = number of data - number of variables or (n - k). Thus the value of F table for the numerator is: 2 - 1 = 1 and the denominator is: 12 - 2 = 10; then obtained value of F from table equal to 4.96.

Fifth: Make a decision

Obtained from the calculation of the F value is 5.158 > F table value of 4.96; then H0 is rejected and H1 accepted. This means that the variable of sales force and number of retailers have significant effect on the sale variable.

Testing with F value can also be done by using curve as follows:

Data Analysis and Its Interpretation: Application in Marketing Research

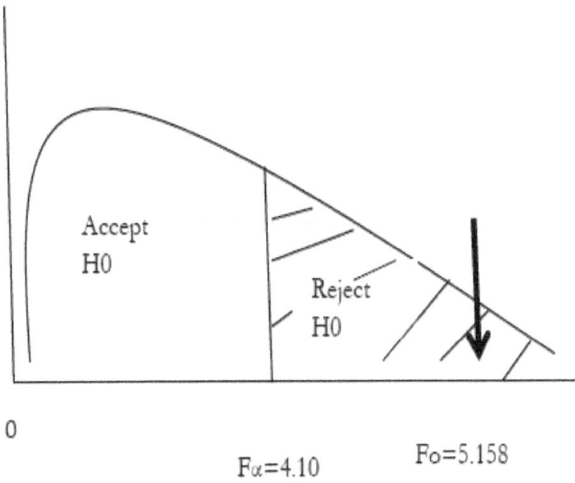

$F\alpha=4.10$ $Fo=5.158$

The curve image shows that the Fo value is in the rejected H0 area so the H1 is accepted. This means that the variable of sales force and number of retailers have a significant effect on the sales variable.

Use the value of significance

First: Determining the hypothesis

H0: $\beta = 0$ (the sales force and number of retailers variables do not significantly affect the sale variable)

H1: $\beta \neq 0$ (the sales force and number of retailers variables significantly affects the sale variable)

Second: Using the following decision-making criteria

- If the value of significance < 0.05; then H0 is rejected and H1 accepted

- If the value of significance > 0.05; then H0 is accepted and H1 is rejected

Third: Make a decision

From the output table above, it is known that the significance value in Sig column is 0.032 < 0.05; then H0 is rejected and H1 accepted. This means that the variable of sales force and number of retailers have a significant effect on the sale variable.

Both using the value of F and the value of significance indicate the same result. In practice, we can use either the F value or the significance value.

Section VI Regression Coefficients

Coefficients[a]

Model		Unstandardized Coefficients		Standardized Coefficients	t	Sig.	Collinearity Statistics	
		B	Std. Error	Beta			Tolerance	VIF
1	(Constant)	63.508	77.068		.824	.431		
	Number of salesforce	13.532	11.519	1.556	1.175	.270	.030	33.895
	Number of retailers	-32.325	50.209	-.853	-.644	.536	.030	33.895

a. Dependent Variable: Sale

This section describes the regression equation to find the constant value, and test the hypothesis of regression coefficient significance:

The regression equation is:

$$Y = a + b_1 X_1 + b_2 X_2$$

Where:

- Y = sale

- X_1 = sales force

- X_2 = number of retailers

- a = constant number of Unstandardized Coefficient which in this research is equal to 63.508. This number is a constant number which means: if there is no additional sales force and retailers (X=0), then sale will be as much as 63.508 (64)

- b1 = the regression coefficient value is 13.532. This number means that in every addition of 1 unit of sales force, then sales will increase by 13.532 (14). Conversely, if this value is negative (-) then sale will decrease by that number.

- b2: the regression coefficient value is -32.325. This number means that in every addition of 1 unit of retailers, then sales will decrease as much as 32.325 (32).

Therefore, the equation becomes:

$Y = 63.508 + 15.226 X_1 - 32.325 X_2$

Hypothesis Testing with Significance Value

To conduct the hypothesis testing using the significance value, the steps are as follows.

First Hypothesis: Sales Force and Sale

First: Determining the hypothesis

H0: the sales force variable does not significantly affect the sale variable

H1: the sales force variable significantly affects the sale variable

Second: Using decision-making criteria

If the value of significance (Sig) < 0.05; then H0 is rejected and H1 accepted

If the value of significance (Sig) > 0.05; then H0 is accepted and H1 is rejected

Third: Make a decision

From the output above, it is known that the significance value in Sig column is 0.270 > 0.05; then H0 is accepted and H1 rejected. This means that the variable of sales force has no a significant effect on the sale variable.

Second Hypothesis: Number of Retailers and Sale

First: Determining the hypothesis

H0: the number of sale variable does not significantly affect the sale variable

H1: the number of sale variable significantly affects the sale variable

Second: Using decision-making criteria

If the value of significance (Sig) < 0.05; then H0 is rejected and H1 accepted

If the value of significance (Sig) > 0.05; then H0 is accepted and H1 is rejected

Third: Make a decision

From the output above, it is known that the significance value in Sig column is 0.536 > 0.05; then H0 is accepted and H1 rejected. This means that the variable of number of retailers has no a significant effect on the sale variable.

Section VII: Residual Statistics V

Residuals Statistics[a]

	Minimum	Maximum	Mean	Std. Deviation	N
Predicted Value	88.32	224.39	147.83	52.603	12
Residual	-82.587	102.675	.000	49.131	12
Std. Predicted Value	-1.131	1.455	.000	1.000	12
Std. Residual	-1.520	1.890	.000	.905	12

a. Dependent Variable: Sale

This section provides an explanation of the minimum value of the predicted sale, namely: 88.32; the maximum value is 224.39; average of the sale is predicted to be 147.83.

Data Analysis and Its Interpretation: Application in Marketing Research

8.3 How to Check Normal Distribution with Kolmogorov and Smirnov

In this section we will use a useful procedure to see whether our data is normally distributed or not. The procedure is called Kolmogorov-Smirnov Goodness of Fit.

Definition

Kolmogorov-Smirnov's Test of Goodness of Fit is used to determine whether the distribution of values in a sample corresponds to a particular theoretical distribution, such as data normality. The data tested is quantitative data that has interval or ratio scale

Steps to Conduct the Analysis

To perform the normality test using Kolmogorov-Smirnov Goodness of Fit Test technique the stages are as follows:

First: Prepare the data to be tested.

The following data is the sales data of certain products on X Supermarket.

Sale
8
7
8
9
7
8
7
8
9
7
7
8
9
9
7
8
7
9
8
7
9
7

7
7
8
8
9
9
7
8

Second: Create the design variable in position Variable View and type

Third: Enter the data in the Data View command

Fourth: Analyze the data in SPSS in the following way:

- Click Analyze > Select Non Parametric Test

- Select Legacy Dialogs > Select 1 Sample KS

- The dialog box appears: Move the variable of sale to be tested to the Test Variable List column

- In the Test Distribution option check the Normal option

- Click Ok

The result will be as below:

One-Sample Kolmogorov-Smirnov Test

		sale
N		30
Normal Parameters[a,b]	Mean	7.87
	Std. Deviation	.819
Most Extreme Differences	Absolute	.255
	Positive	.255
	Negative	-.183
Test Statistic		.255
Asymp. Sig. (2-tailed)		.000[c]

a. Test distribution is Normal.
b. Calculated from data.
c. Lilliefors Significance Correction.

Fifth: Make a hypothesis

H0: Data is normally distributed

H1: Data is not normally distributed

Sixth: Determining the level of probability / significance, namely 0.05

Seventh: Determining the criteria for hypothesis testing

If Asymp sig < 0.05 H0 is rejected, H1 is accepted

If Asymp sig > 0.05 H0 is accepted, H1 is rejected

Eighth: Make a decision

As seen from the output above Asymp Sig value derived from data is $0.000 < 0.05$, then H0 is rejected and H1 accepted.

Ninth: Make a conclusion

Because H0 is rejected and H1 is accepted then the data is not normally distributed. If the data is not normally distributed we should not use linear regression procedures. Because in the linear regression the link identity between the independent variable and the dependent variable is the normal data distribution.

8.4 Linearity Test and Heteroscedasticity with Curves

In the next example, we do linearity and heteroscedasticity testing by using promotion and sales variables.

Promotion	Sale
30	200
25	140
15	130
24	175
31	240

27	120
10	160
9	90
29	300
12	89
16	70
14	60

To analyze the data above use the following steps:

• Analyze > Regression > Linear

• Move the promotional variable to the Independent column (s) and sale variable to the Dependent column

• Click Plot

• Select SDRESID and move to column Y; select ZPRED and move to the X column then click Next

• Select select ZPRED and move to the Y column; select DEPENDNT and move to the X column; check (v) in the Histogram and Normal Probability Plot options then Click Continue

• Click OK

The result will be as below.

Data Analysis and Its Interpretation: Application in Marketing Research

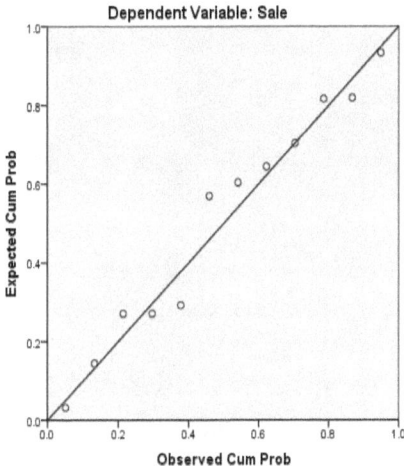

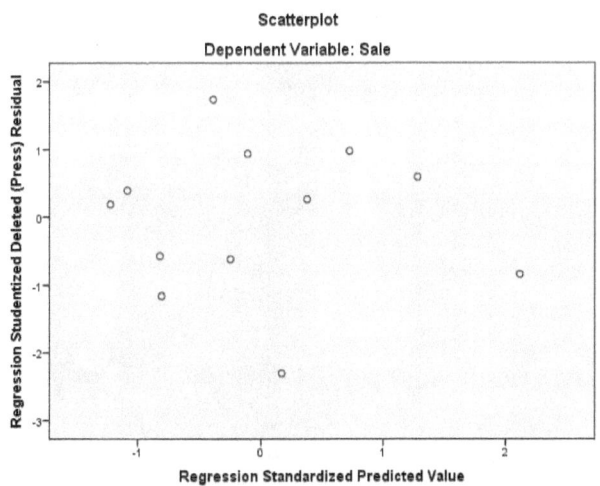

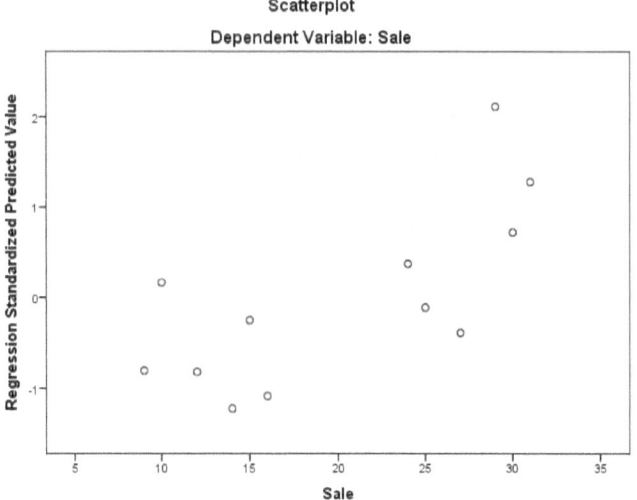

8.4.1 Viewing the Linearity Relationship between Promotion and Sale Variables

The graph results above can be used to check linearity as follows

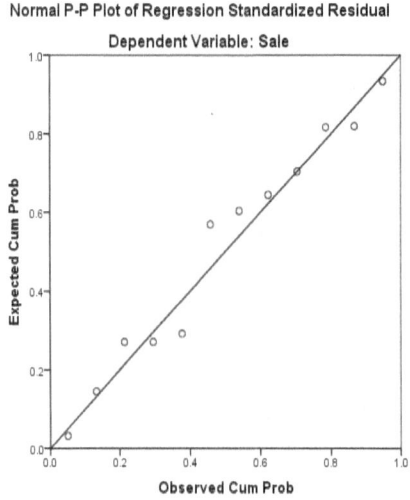

If seen above the output shows the distribution of data from the lower left to the upper right indicating a positive linear relationship between the

Data Analysis and Its Interpretation: Application in Marketing Research

independent variable of promotion and sale.

8.4.2 Viewing Heteroscedasticity with Residual Curve and Prediction Value

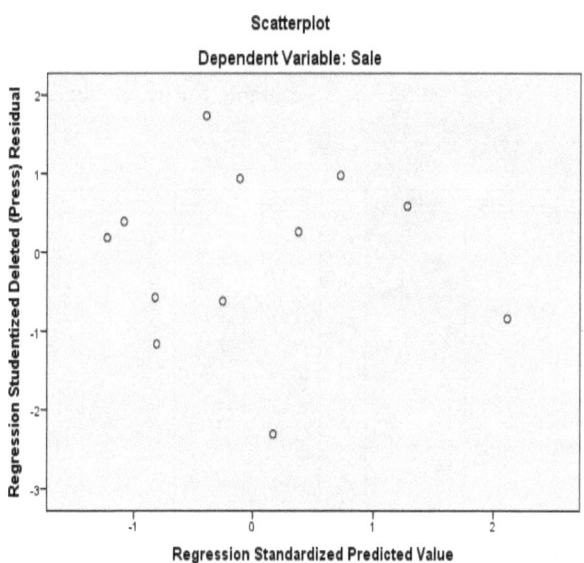

If in the output of the residual curve and the predicted value of the data distribution do not form certain patterns such as an S curve, a U curve or a horizontal curve; then the distribution of data is called random and does not form certain patterns. Thus in data distribution there is no occurrence of heteroscedastisity. In the curve above the graph of the data distribution does not form certain patterns. Thus there is no heteroscedasticity in the data being analyzed.

8.5 How to See Multicolinearity in Multiple Linear Regression

In the relationship between independent variables will occur multicollinearity when the value of correlation coefficient between independent variables in multiple linear regression has a very high value or very low. The values used to test multicollinearity include:

• Value of variance inflation factor (VIF) with provision if VIF value is > 5 then multicollinearity occurs;

- Correlation value between the independent variables is ≥ 0.9 then the multicollinearity occurs (Hair, 2010)

8.5.1 Case

The following is the data we will use as an example for multicollinearity testing.

No	Product	Service	Complaint	Satisfaction	No	Product	Service	Complaint	Satisfaction
1	18	17	16	15	51	13	14	13	12
2	15	17	18	12	52	13	13	14	10
3	17	14	16	14	53	15	16	16	15
4	14	14	14	13	54	12	12	14	12
5	15	15	16	12	55	13	12	15	9
6	17	15	16	13	56	15	15	16	14
7	13	16	12	14	57	14	18	12	10
8	19	19	20	11	58	13	11	13	11
9	15	16	17	14	59	10	11	13	12
10	19	19	18	14	60	16	18	16	12
11	14	16	17	12	61	14	11	11	10
12	15	11	15	10	62	17	12	12	13
13	14	14	13	12	63	16	14	13	12
14	16	16	18	11	64	14	15	12	10
15	10	14	16	12	65	13	11	14	10
16	13	15	17	13	66	13	11	14	11
17	19	12	17	14	67	12	11	14	9
18	15	12	16	14	68	13	14	13	10
19	15	12	14	15	69	15	12	13	9
20	16	14	15	14	70	13	13	17	12
21	12	11	15	14	71	13	14	12	11
22	18	16	14	14	72	13	16	15	12
23	13	15	15	11	73	15	14	15	12
24	14	14	13	12	74	14	13	15	10
25	13	13	17	11	75	14	14	13	10
26	10	10	12	9	76	16	16	18	15
27	13	17	13	9	77	20	19	22	17
28	12	12	11	10	78	17	20	20	16
29	11	14	12	8	79	16	20	20	15
30	9	14	8	7	80	18	16	16	16
31	10	10	13	12	81	15	18	19	12
32	8	12	12	9	82	16	19	19	12
33	13	14	12	10	83	20	16	15	13
34	8	15	14	9	84	21	22	22	19
35	11	8	12	10	85	19	19	19	12
36	13	13	16	11	86	18	15	16	13
37	11	14	11	9	87	18	19	20	12
38	9	15	14	11	88	17	15	18	12
39	15	11	12	12	89	21	17	20	17
40	11	12	15	8	90	16	17	19	17
41	11	12	13	8	91	16	19	18	16
42	9	14	14	9	92	19	16	20	16
43	10	17	11	9	93	17	19	18	16
44	9	10	12	8	94	19	18	21	13
45	12	9	12	7	95	21	17	20	19
46	13	15	15	8	96	19	16	17	15
47	13	8	14	10	97	19	15	19	15
48	14	11	11	9	98	17	19	19	15
49	11	13	10	9	99	19	19	19	12
50	12	13	15	13	100	18	17	19	14

Data Analysis and Its Interpretation: Application in Marketing Research

8.5.3 How to Conduct Analysis in SPSS

To analyze the way is:

• Select Analyze > Regression > Linear

• Enter the Satisfaction variable into the Dependent column

• Enter Product Quality, Service Quality and Complaint Handling variables into Independent column (s)

• Click on Statistics; do a check on the options Descriptives and Collinearity Diagnostics then click Continue

• Click OK

Looking at Multicollinearity with Correlation Approach

The table below is the output of correlation that we will use to assess the occurrence of multicollinearity.

Correlations

		Customer Satisfaction	Product Quality	Service Quality	Complaint Handling
Pearson Correlation	Customer Satisfaction	1.000	.730	.548	.702
	Product Quality	.730	1.000	.600	.724
	Service Quality	.548	.600	1.000	.684
	Complaint Handling	.702	.724	.684	1.000
Sig. (1-tailed)	Customer Satisfaction	.	.000	.000	.000
	Product Quality	.000	.	.000	.000
	Service Quality	.000	.000	.	.000
	Complaint Handling	.000	.000	.000	.
N	Customer Satisfaction	100	100	100	100
	Product Quality	100	100	100	100
	Service Quality	100	100	100	100
	Complaint Handling	100	100	100	100

When viewed on the output above the value of correlation coefficients among the three independent variables are as follows:

- The correlation value between the independent variable of product quality and service quality is 0.600

- The correlation value between product quality and complaint handling is 0.724

- The correlation value between service quality and complaint handling is 0.684

Since the correlation values among the independent variables are smaller than 0.9; then there is no multicollinearity in the data above.

8.5.4 Looking at Multicollinearity with the Variance Inflation Factor (VIF)

The following is the output of VIF.

Collinearity Statistics	
Tolerance	VIF
.455	2.199
.509	1.963
.379	2.641

From the above output it is seen that the VIF values for the three independent variables of product quality, service quality and complaint handling are less than 5; then there is no multicollinearity in the data above.

8.6 How to Assess Data from the Occurrence of Autocorrelation Using Durbin Watson Test

Autocorrelation has a meaning in one variable there are values that correlate each other. To see whether in the data in one variable there is autocorrelation or not we can use Durbin – Watson (DW) test. The range of DW values is from 0 - 4. There is no autocorrelation if the DW value: $-2 \leq DW \leq 2$ (Anderson, 2001: 733).

8.6.1 Case

Using the following data to perform the autocorrelation test according to

Data Analysis and Its Interpretation: Application in Marketing Research

Durbin – Watson

No	Product	Service	Complaint	Satisfaction	No	Product	Service	Complaint	Satisfaction
1	18	17	16	15	51	13	14	13	12
2	15	17	18	12	52	13	13	14	10
3	17	14	16	14	53	15	16	16	15
4	14	14	14	13	54	12	12	14	12
5	15	15	16	12	55	13	12	15	9
6	17	15	16	13	56	15	15	16	14
7	13	16	12	14	57	14	18	12	10
8	19	19	20	11	58	13	11	13	11
9	15	16	17	14	59	10	11	13	12
10	19	19	18	14	60	16	18	16	12
11	14	16	17	12	61	14	11	11	10
12	15	11	15	10	62	17	12	12	13
13	14	14	13	12	63	16	14	13	12
14	16	16	18	11	64	14	15	12	10
15	10	14	16	12	65	13	11	14	10
16	13	15	17	13	66	13	11	14	11
17	19	12	17	14	67	12	11	14	9
18	15	12	16	14	68	13	14	13	10
19	15	12	14	15	69	15	12	13	9
20	16	14	15	14	70	13	13	17	12
21	12	11	15	14	71	13	14	12	11
22	18	16	14	14	72	13	16	15	12
23	13	15	15	11	73	15	14	15	12
24	14	14	13	12	74	14	13	15	10
25	13	13	17	11	75	14	14	13	10
26	10	10	12	9	76	16	16	18	15
27	13	17	13	9	77	20	19	22	17
28	12	12	11	10	78	17	20	20	16
29	11	14	12	8	79	16	20	20	15
30	9	14	8	7	80	18	16	16	16
31	10	10	13	12	81	15	18	19	12
32	8	12	12	9	82	16	19	19	12
33	13	14	12	10	83	20	16	15	13
34	8	15	14	9	84	21	22	22	19
35	11	8	12	10	85	19	19	19	12
36	13	13	16	11	86	18	15	16	13
37	11	14	11	9	87	18	19	20	12
38	9	15	14	11	88	17	15	18	12
39	15	11	12	12	89	21	17	20	17
40	11	12	15	8	90	16	17	19	17
41	11	12	13	8	91	16	19	18	16
42	9	14	14	9	92	19	16	20	16
43	10	17	11	9	93	17	19	18	16
44	9	10	12	8	94	19	18	21	13
45	12	9	12	7	95	21	17	20	19
46	13	15	15	8	96	19	16	17	15
47	13	8	14	10	97	19	15	19	15
48	14	11	11	9	98	17	19	19	15
49	11	13	10	9	99	19	19	19	12
50	12	13	15	13	100	18	17	19	14

8.6.3 How to Analyze in SPSS

How to do the analysis is as follows.

• Click Analyze > Regression > Linear Move the satisfaction variable to the Dependent column

- Move product quality, service quality and complaint handling variables to Independent column

- Select Statistics: Check (v) in the Durbin - Watson option. Press Continue

- Click Ok

The output that we will use lies in the Summary Model table as shown below.

Model Summary^b

Model	R	R Square	Adjusted R Square	Std. Error of the Estimate	Durbin-Watson
1	.773^a	.597	.584	1.685	1.911

a. Predictors: (Constant), Complaint Handling, Service Quality, Product Quality
b. Dependent Variable: Customer Satisfaction

At the output above the DW value is 1.911 thus the value is less than 2; then the conclusion there is no autocorrelation.

8.7 How to Assess Heteroscedasticity or Variance Differences

To understand the meaning of heteroscedasticity is necessary to understand in advance the notion of homoscedasticity. Homoscedasticity is the description of data where the error terms (e) appear to be constant beyond the reach of certain independent variable values. The assumption of the equality of the population error variance of ε (where ε is estimated from the sample value) is critical if applied to the correct linear regression. When the error limit has an increasing variance, then the data is called heteroscedasticity. In other words, homoscedasticity is the assumption that the dependent variable shows the same level of variance for all the independent variables. If the spread of the variance value on all independent variables is not the same then the relationship is said to be heteroscedasticity. To test homoscedasticity (the occurrence of variance equality in all independent variables) Levene testing is used for the metric and non metric variables. Variance equality occurs when the value of significance (sig) in Levene test > 0.05. Levene testing can be explained by making the null hypothesis (H0) which states that "there is variance equality on all independent variables" and alternative hypothesis (H1) which states " there is no variance equality on all independent variables ". The provision of hypothesis testing is based on the value of significance: if sig value > 0.05 H0 is accepted; if significance value < 0.05 H0 is rejected. If the metric scale variables we can use the test Box's M. Terms of testing is the same as the test using Levene test. To find out whether there is heteroscedastisity in the above test can be known from the significance value. If the significance value (sig) < 0.05, then in the model heteroscedastisity occurs. Similarly, it applies to the Glejser test if the regression coefficient in the independent variable under study yields a value greater than 0.05; then there is no heteroscedasticity.

8.7.2 Seeing Heteroscedasticity with Glejser Test

Case: The following is the data we will analyze whether there is heteroscedasticity or not by using the Glejser approach.

Y	X1	X2	X3
.533	.333	3.000	91.940
.607	.333	3.000	58.410
.533	.500	3.000	36.590
.664	.333	4.000	43.860
.467	.600	3.000	44.470
.673	.300	4.000	50.110
.729	.400	3.000	80.000
.551	.667	3.000	52.500
.570	.333	3.000	26.420
.598	.333	5.000	55.530
.495	.333	3.000	51.180
.607	.333	3.000	52.200
.617	.333	3.000	69.290
.551	.500	3.000	98.180
.748	.375	3.000	80.530
.551	.429	3.000	71.490
.729	.500	3.000	80.660
.458	.333	3.000	97.820
.757	.375	3.000	50.070
.701	.429	3.000	51.000
.561	.667	3.000	52.570
.579	.500	3.000	35.700
.822	.400	3.000	90.020
.570	.333	3.000	49.680
.533	.333	3.000	17.860
.804	.333	3.000	12.000
.561	.500	3.000	75.700
.570	.333	3.000	43.620
.897	.333	3.000	10.170
.495	.333	3.000	25.550
.495	.429	3.000	76.240
.542	.500	3.000	58.770
.458	.333	3.000	40.970
.729	.333	3.000	31.500
.542	.333	3.000	36.000
.720	.667	3.000	15.560
.991	.333	4.000	51.010
.673	.333	3.000	58.130
.607	.600	3.000	44.600
.673	.375	3.000	35.000
.542	.333	3.000	40.990
.523	.333	3.000	24.630
.598	.400	4.000	60.840
.505	.333	3.000	33.780
.682	.333	3.000	38.770
.626	.333	3.000	35.600
.551	.500	3.000	35.900
.729	.800	3.000	85.000
.701	.400	3.000	46.530
.561	.333	3.000	89.470

Data Analysis and Its Interpretation: Application in Marketing Research

8.7.3 How to Analyze in IBM SPSS the steps are as follows:

• Analyze > Regression > Linear

• Move X1, X2 and X3 variables to Independent Variable column and Y to Dependent Variable column

• Click Save

• In the Residual section check on the Unstandardized > Continue option

• Click OK

Ignore the output, then we see in the Data View appears a new variable RES_1

Then the next step is to create the RES_2 variable in the following way:

• Transform > Compute Variable in the Target Variable field type RES_2

• In the Numeric Expression field type ABS_RES (RES_1)

• Click OK

Ignore the output. Then we see in the Data View appears a new variable RES_2 as follows

Then we proceed to the next process do the analysis for the new variable in the following way.

• Analyze > Regression > Linear

• Replace the Y dependent variable by moving it back to the left column and enter the RES_2 variable into the Dependent Variable column

• Click Save

• Then uncheck (v) in the Residual section of the Unstandardized > Continue option

• Click OK

Output as follows:

Coefficients[a]

Model		Unstandardized Coefficients		Standardized Coefficients	t	Sig.
		B	Std. Error	Beta		
1	(Constant)	.057	.093		.613	.543
	X1	-.014	.086	-.025	-.166	.869
	X2	.015	.025	.089	.598	.553
	X3	.000	.000	-.061	-.410	.684

a. Dependent Variable: RES_2

The output we use is the output in the Table Coefficients section above. There is no heteroscedasticity in the above data because the significance values for the X1, X2 and X3 are more than 0.05. Thus we can conclude that all variables are free from heteroscedasticity.

8.8 Overcoming Multicollinearity in Multiple Linear Regression

8.8.1 Case

The case that will be used is the automotive industry group that tracks sales of various private motor vehicles. In an attempt to identify a model that can be practiced or not applicable, this study attempts to build a relationship between vehicle sale and vehicle characteristics. For this purpose we will use the IBM SPSS file with the name *car_sales.sav*. In this case, the dependent variable is *sales in thousands* and the independent variables (predictors) is *fuel efficiency, length, price in thousand, vehicle type, width, engine size, wheel base, curb weight* , and *horse power*.

Steps to analysis are as follows
First: Open the car_sales.sav file and then conduct the analysis by using the following commands
- **Analyze > Regression > Linear**
- In **the Dependent** column enter the variable **sales in thousands**
- In the **Independent** column enter the variables of **fuel efficiency, length, price in thousand, vehicle type, width, engine size, wheel base, curb weight , and horse power** .
- Select **Statistics:** check options **estimates, fit models, descriptive and collinearity diagnostics**. For

# Data Analysis and Its Interpretation: Application in Marketing Research

the **Residuals** option , check **Durbin - Watson** and **Case Wise Diagnostics**, check **all cases** . Click **Continue**
- Select **Plot**: check on **option produce all partial plots, histogram and normal probability plot**. Click **Continue**
- Click **OK**

The results will be as below:
In this section not all results and interpretations will be discussed.

Part 1: Processed Variables

Variables Entered/Removed[a]

Model	Variables Entered	Variables Removed	Method
1	Horsepower, Vehicle type, Length, Width, Fuel efficiency, Price in thousands, Wheelbase, Engine size, Curb weight[b]	.	Enter

a. Dependent Variable: Sales in thousands
b. All requested variables entered.

The above output provides information to us that there are 9 independent variables included as predictors , namely: *fuel efficiency, length, price in thousand, vehicle type, width, engine size, wheel base, curb weight* , and *horse power* and 1 variable (dependent) as the target, that is *sales in thousands* .

Part 2: The Summary Model

Model Summary[b]

Model	R	R Square	Adjusted R Square	Std. Error of the Estimate	Durbin-Watson
1	.582[a]	.338	.296	57.824438	1.282

a. Predictors: (Constant), Horsepower, Vehicle type, Length, Width, Fuel efficiency, Price in thousands, Wheelbase, Engine size, Curb weight
b. Dependent Variable: Sales in thousands

In the above output, the important values are:
- The value of Rsquare (R^2) is 0.338 has the meaning that the variance proportion of the value of *'sales in thousands'* can be explained by using the above 9 predictors as much as 0.338 or 33.8%. This magnitude value has not shown good result yet. Ideally the value of R^2 is at least greater than

0.5 (more than 50%).
- Adjusted R² Value: Adjusted R² value is 0.296 indicating that the suitability of the regression model we are making is not so true because this value is far below 1. The model fit is measured by this value with the provision that the value close to 1 has matching the model fit.
- Durbin - Watson (DW) Value: The DW value of 1.282 indicates no autocorrelation in the regression model we make.

Section 3: ANOVA Table

ANOVA^a

Model		Sum of Squares	df	Mean Square	F	Sig.
1	Regression	242827.014	9	26980.779	8.069	.000^b
	Residual	474800.520	142	3343.666		
	Total	717627.535	151			

a. Dependent Variable: Sales in thousands

b. Predictors: (Constant), Horsepower, Vehicle type, Length, Width, Fuel efficiency, Price in thousands, Wheelbase, Engine size, Curb weight

In this section shows the feasibility of the regression model we create. To test the feasibility we do hypothesis testing with steps - steps as follows:

Formulate Hypotheses
- H0: There is no linear relationship between variables of *sales in thousands* with *fuel efficiency, length, price in thousand, vehicle type, width, engine size, wheel base, curb weight* , and*horse power*
- H1: There is a linear relationship between variable *sales in thousands* with *fuel efficiency, length, price in thousand, vehicle type, width, engine size, wheel base, curb weight* , and*horse power*

Testing Criteria
- If significance (sig) < 0.05 H0 is rejected
- If significance (sig)> 0.05 H0 is accepted

Decision
With significance value on the ANOVA table shows the value of Sig 0.000 < 0.05; then H0 is rejected and H1 accepted. Thus, there is a linear relationship between variables of *sales in thousands* with *fuel efficiency, length, price in thousand, vehicle type, width, engine size, wheel base, curb weight* , and *horse power*. Thus, the regression model can be said to be feasible .

Data Analysis and Its Interpretation: Application in Marketing Research

Part 4: Regression Coefficient

Coefficients[a]

Model		Unstandardized Coefficients		Standardized Coefficients	t	Sig.	Collinearity Statistics	
		B	Std. Error	Beta			Tolerance	VIF
1	(Constant)	-346.436	160.097		-2.164	.032		
	Fuel efficiency	.788	2.276	.049	.346	.730	.231	4.334
	Length	.750	.827	.146	.907	.366	.179	5.599
	Price in thousands	.054	.745	.011	.073	.942	.192	5.215
	Vehicle type	43.318	18.749	.278	2.310	.022	.323	3.099
	Width	.753	2.416	.038	.312	.756	.316	3.164
	Engine size	31.802	11.124	.484	2.859	.005	.162	6.159
	Wheelbase	3.639	1.347	.407	2.702	.008	.205	4.880
	Curb weight	-66.370	19.366	-.613	-3.427	.001	.146	6.864
	Horsepower	-.484	.246	-.399	-1.965	.051	.113	8.847

a. Dependent Variable: Sales in thousands

This section of the menu shows the value of the regression coefficient indicating a linear or no relationship between each predictor and the dependent variable. To see the relationship we can do hypothesis testing one by one as below:

Formulate Hypothesis for Fuel Efficiency
- H0: There is no linear relationship between variable *sales in thousands* with *fuel efficiency*,
- H1: There is a linear relationship between variable *sales in thousands* with *fuel efficiency*

Testing Criteria
- If significance (sig) <0.05 H0 is rejected
- If significance (sig) > 0.05 H0 is accepted

Decision
Based on the calculation above the significance (Sig) value for 0.730 > 0.05; then H0 is accepted and H1 is rejected. Thus, there is no linear relationship between variables of *sales in thousands* with *fuel efficiency*.

Formulate Hypothesis for Length
- H0: There is no linear relationship between variable *sales in thousands* with *length* ,
- H1: There is a linear relationship between variable *sales in thousands* with *length*

Testing Criteria
- If significance (sig) <0.05 H0 is rejected

- If significance (sig) > 0.05 H0 is accepted

Decision
The significance (Sig) value is 0.366 > 0.05; then H0 is accepted and H1 is rejected. Thus is, there is no linear relationship between variables of *sales in thousands* with *length*.

Formulate Hypothesis for Price
- H0: There is no linear relationship between variables of *sales in thousands* with *price*
- H1: There is a linear relationship between variables of *sales in thousands* with *price*

Testing Criteria
- If significance (sig) < 0.05 H0 is rejected
- If significance (sig) > 0.05 H0 is accepted

Decision
The significance value is 0.942 > 0.05; then H0 is accepted and H1 is rejected. Thus, there is no linear relationship between variables of *sales in thousands* with *price*.

Formulate Hypotheses for Vehicle type
- H0: There is no linear relationship between variables of *sales in thousands* with *vehicle type*
- H1: There is a linear relationship between variables of *sales in thousands* with *vehicle type*

Testing Criteria
- If significance (sig) <0.05 H0 is rejected
- If significance (sig) > 0.05 H0 is accepted

Decision
The significance value shows 0.022 < 0.05; then H0 is rejected and H1 accepted. Thus there is linear relationship between variables of *sales in thousands* with *vehicle type*.

Formulate Hypotheses for Width
- H0: There is no linear relationship between variables of *sales in thousands* with *width*
- H1: There is a linear relationship between variables of *sales in thousands* with *width*

Testing Criteria
- If significance (sig) <0.05 H0 is rejected
- If significance (sig)> 0.05 H0 is accepted

Decision
The significance value shows as 0.756 > 0.05; then H0 is accepted and H1 is rejected. This means there is no linear relationship between variables of *sales in thousands* with *width*.

Formulate Hypotheses for Engine size
- H0: There is no linear relationship between variables of *sales in thousands* with *engine size*
- H1: There is a linear relationship between variables of *sales in thousands* with *engine size*

Testing Criteria
- If significance (sig) < 0.05 H0 is rejected
- If significance (sig) > 0.05 H0 is accepted

Decision
The significance value shows 0.005 < 0.05; then H0 is rejected and H1 accepted. Thus there is linear relationship between variables of *sales in thousands* with *engine size*.

Formulate Hypothesis for Wheelbase
- H0: There is no linear relationship between variables of *sales in thousands* with *wheelbase*
- H1: There is a linear relationship between variables of *sales in thousands* with *wheelbase*

Testing Criteria
- If significance (sig) < 0.05 H0 is rejected
- If significance (sig) > 0.05 H0 is accepted

Decision
The value of significance as much as 0.008 < 0.05; then H0 is rejected and H1 accepted. Thus there is linear relationship between variables of *sales in thousands* with *wheelbase*.

Formulate Hypotheses for Curbweight
- H0: There is no linear relationship between variables of *sales in thousands* with *curbweight*
- H1: There is a linear relationship between variables of *sales in thousands* with *curbweight*

Testing Criteria
- If significance (sig) < 0.05 H0 is rejected
- If significance (sig) > 0.05 H0 is accepted

Decision

The value of significance shows 0.001 < 0.05; then H0 is rejected and H1 accepted. Thus there is linear relationship between variables of *sales in thousands* with *curbweight*.

Formulate hypothesis for Horsepower
- H0: There is no linear relationship between variables of *sales in thousands* with *horsepower*
- H1: There is a linear relationship between variables of *sales in thousands* with *horsepower*

Testing Criteria
- If significance (sig) < 0.05 H0 is rejected
- If significance (sig) > 0.05 H0 is accepted

Decision

The significance value shows as 0.051 > 0.05; then H0 is accepted and H1 is rejected. Thus there is no linear relationship between variables of *sales in thousands* with *horsepower*.

In addition to the above values, we will also detect the collinearity by using the VIF value contained in the *Collinearity Statistics* column The VIF value > 5 indicating multicollinearity occurs. It occurs variables of *length* as much as 5.999; *engine size* as much as 6.159; *curbwhite* as much as 6.684 and *horsepower* as much as 8.847.

Section 4: Diagnosis of Collinearity with Condition Index value

Collinearity Diagnostics[a]

Model	Dimension	Eigenvalue	Condition Index	Variance Proportions								
				(Constant)	Fuel efficiency	Length	Price in thousands	Vehicle type	Width	Engine size	Wheelbase	Curb weight	Horsepower
1	1	8.936	1.000	.00	.00	.00	.00	.00	.00	.00	.00	.00	.00
	2	.730	3.499	.00	.00	.00	.00	.30	.00	.00	.00	.00	.00
	3	.259	5.878	.00	.01	.00	.07	.00	.00	.01	.00	.00	.00
	4	.050	13.432	.00	.02	.00	.31	.10	.00	.15	.00	.00	.01
	5	.013	26.358	.00	.13	.00	.02	.25	.00	.07	.00	.16	.13
	6	.008	34.255	.00	.08	.00	.25	.00	.00	.68	.00	.00	.67
	7	.003	55.303	.01	.39	.03	.23	.00	.01	.03	.06	.76	.16
	8	.002	72.298	.18	.27	.07	.02	.07	.12	.00	.11	.00	.01
	9	.001	123.672	.07	.01	.81	.03	.11	.08	.00	.66	.02	.01
	10	.000	140.649	.75	.08	.09	.07	.16	.79	.05	.16	.06	.00

a. Dependent Variable: Sales in thousands

To make a diagnosis whether there is a problem of collinearity or not we can use the Condition Index Value (CIV) and Eigenvalue. If we use the CIV the value should be less than 15. The above output on the Condition Index Value indicates the problem of collinearity on the independent variables : 4 (vehicle type), 5 (width), 6 (engine size), 7(wheelbase), 8 (curbweight), and 9 (horse power). Beside that value, we can also use the Eigenvalue. Variables that have a problem of colinearity is also called multicollinearity.

Data Analysis and Its Interpretation: Application in Marketing Research

8.8.2 Solving Collinearity Problems
In this part, we will correct this problem by doing the analysis as follows:
- If not already done by IBM SPSS, change all independent variables used as predictors to Z value in the following way:
  - **Analyze > Descriptive Statistics > Descriptive**
  - Move all the independent variables that become predictors to Variable (s) column
  - Check the **Save Standardized Values as Variable** option.
  - Click **Ok**

- If there is already a variable with the initial z, ignore the above command and re-do the regression analysis with the following command:
  - **Analyze > Regression > Linear**
  - In the **Dependent** column enter the **Log Transformed sales** variable **[insales]**
  - In the column of **Independent (s)** enter the variables that have been changed to the value Z. The variables begin with zscore and then followed the name of the variable **zscore: fuel efficiency, zscore: length, zscore: width, zscore: engine size, zscore: wheel base, zscore: curb weight**, and **zscore: horse power**.
  - Select **Statistics** : check options **estimates, fit models, descriptive** and **collinearity diagnostics**. For the **Residuals** option, check **Durbin - Watson** and **Case Wise Diagnostics**, check **all cases**. Click **Continue**
  - Select **Plot**: check on option **produce all partial plots, histogram** and **normal probability plot**. Click **Continue**
  - Click **OK**

The results and interpretations are as follows:
Part 1 : Summary Model

Model Summary[b]

Model	R	R Square	Adjusted R Square	Std. Error of the Estimate	Durbin-Watson
1	.615[a]	.378	.349	1.06909	1.531

a. Predictors: (Constant), Zscore: Horsepower, Zscore: Wheelbase, Zscore: Fuel capacity, Zscore: Width, Zscore: Length, Zscore: Engine size, Zscore: Curb weight

b. Dependent Variable: Log-transformed sales

If we look at the value of R^2 in this second analysis it has increased from 0.338 to 0.378. It has meaning that the correction has been done contributes a positive impact on increasing the value of R^2. Similarly, the value of *Adjusted* R^2 is increased increased from 0.296 to 0.349. Thus the fit model in this regression is getting better.

Part 2: ANOVA Table

ANOVA[a]

Model		Sum of Squares	df	Mean Square	F	Sig.
1	Regression	102.319	7	14.617	12.789	.000[b]
	Residual	168.015	147	1.143		
	Total	270.334	154			

a. Dependent Variable: Log-transformed sales

b. Predictors: (Constant), Zscore: Horsepower, Zscore: Wheelbase, Zscore: Fuel capacity, Zscore: Width, Zscore: Length, Zscore: Engine size, Zscore: Curb weight

In this second ANOVA table output the significance value in the Sig column remains 0.00 as the first output thus the conclusion we take remains the same, i.e. the regression model is feasible. But if we see the value of F then on the second output of this value of F higher than the original at 8.069 to 12.789. This value means that the regression model is getting better because the higher F value indicates the better model.

Part 3: Collinearity Diagnostics

Collinearity Diagnostics[a]

				Variance Proportions							
Model	Dimension	Eigenvalue	Condition Index	(Constant)	Zscore: Fuel capacity	Zscore: Length	Zscore: Width	Zscore: Engine size	Zscore: Wheelbase	Zscore: Curb weight	Zscore: Horsepower
1	1	4.808	1.000	.00	.01	.01	.01	.01	.01	.01	.01
	2	1.032	2.158	.02	.00	.04	.00	.03	.06	.00	.10
	3	.999	2.193	.98	.00	.00	.00	.00	.00	.00	.00
	4	.497	3.110	.00	.21	.09	.03	.01	.00	.06	.06
	5	.282	4.126	.00	.00	.07	.89	.00	.06	.00	.07
	6	.142	5.818	.00	.07	.44	.02	.01	.55	.31	.17
	7	.128	6.136	.00	.04	.06	.04	.91	.06	.01	.58
	8	.112	6.561	.00	.67	.28	.00	.04	.26	.60	.02

a. Dependent Variable: Log-transformed sales

After the second analysis we see that the Condition Index value on all the independent variables used as predictors fall below 15. This decrease has no meaning to multicollinearity occurence in the regression model that we have created based on the Condition Index value. However, if seen from the VIF value, there is no change in the problem of collinearity because some variables still have values > 5, namely *length, price, engine size, curbweight* and *horsepower*.

Data Analysis and Its Interpretation: Application in Marketing Research

Coefficients[a]

Model		Unstandardized Coefficients		Standardized Coefficients	t	Sig.	Collinearity Statistics	
		B	Std. Error	Beta			Tolerance	VIF
1	(Constant)	3.293	.086		38.337	.000		
	Zscore: Fuel capacity	-.210	.180	-.159	-1.171	.244	.229	4.372
	Zscore: Length	.172	.174	.130	.990	.324	.245	4.081
	Zscore: Width	-.064	.150	-.048	-.424	.672	.331	3.020
	Zscore: Engine size	.607	.201	.456	3.024	.003	.186	5.381
	Zscore: Wheelbase	.486	.182	.368	2.668	.008	.222	4.496
	Zscore: Curb weight	-.080	.205	-.061	-.391	.696	.176	5.678
	Zscore: Horsepower	-1.037	.162	-.779	-6.379	.000	.284	3.523

a. Dependent Variable: Log-transformed sales

To solve this case we do the analysis with the factor analysis procedure as below:

- **Analyze> Factor Analysis> Dimension Reduction**

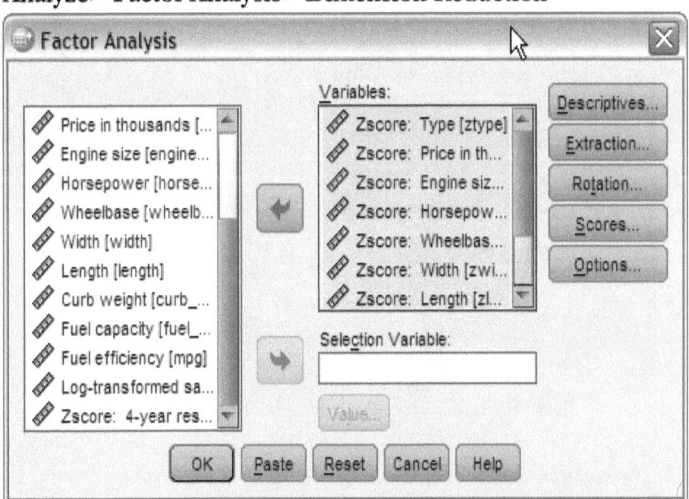

- Move the 9 variables with Zscore starting from zmpg to zhorsepower .
- Select Extraction

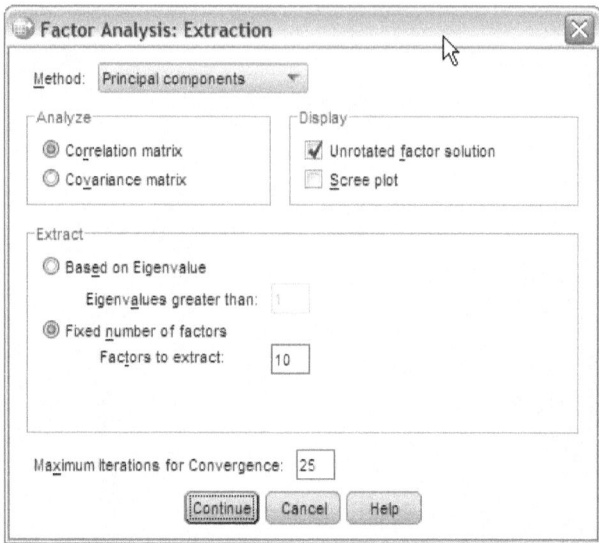

- Check **Fixed number of factors**, on **Factors to extract** fill the value of 9 > **Continue**
- Click **Rotation**

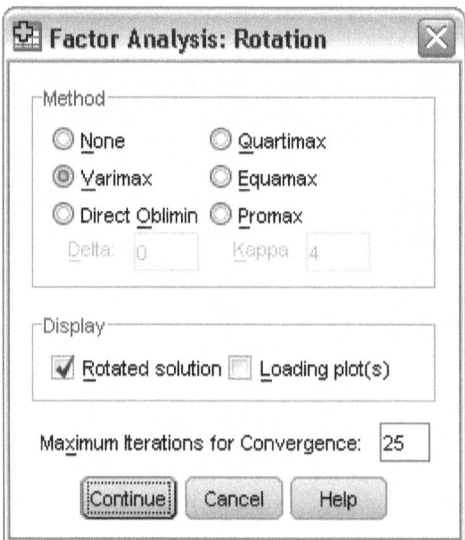

- In the **Method** section check the options **Varmax > Continue**
- Click **Scores**

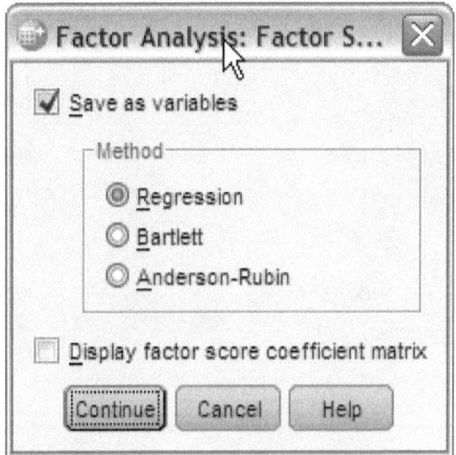

- Check on the **Save as variables > Continue**
- Click **Ok**

After the above process, repeat the regression analysis by entering new variables with the prefix REGR Factor Score to replace all variables with the Zscore prefix as follows:

- **Analyze > Regression > Linear**

- Move the **Log-transformed sale [insales]** variable to the **Dependent** column.
- Move the 9 variables beginning with **REGR factor score** into **Independent** column.
- Click **OK**

181

The results and interpretation is as follows:

Model Summary[b]

Model	R	R Square	Adjusted R Square	Std. Error of the Estimate	Durbin-Watson
1	.688[a]	.473	.440	.99765	1.564

a. Predictors: (Constant), REGR factor score 9 for analysis 1, REGR factor score 2 for analysis 1, REGR factor score 7 for analysis 1, REGR factor score 5 for analysis 1, REGR factor score 3 for analysis 1, REGR factor score 6 for analysis 1, REGR factor score 1 for analysis 1, REGR factor score 4 for analysis 1, REGR factor score 8 for analysis 1

b. Dependent Variable: Log-transformed sales

The value of R^2 permanent as originally on analysis second, that is 0.480.

ANOVA[a]

Model		Sum of Squares	df	Mean Square	F	Sig.
1	Regression	127.051	9	14.117	14.183	.000[b]
	Residual	141.332	142	.995		
	Total	268.383	151			

a. Dependent Variable: Log-transformed sales

b. Predictors: (Constant), REGR factor score 9 for analysis 1, REGR factor score 2 for analysis 1, REGR factor score 7 for analysis 1, REGR factor score 5 for analysis 1, REGR factor score 3 for analysis 1, REGR factor score 6 for analysis 1, REGR factor score 1 for analysis 1, REGR factor score 4 for analysis 1, REGR factor score 8 for analysis 1

The value of significance on Sig column at ANOVA table is 0.000. This means the model is a a good fit.

Coefficients^a

Model		Unstandardized Coefficients B	Std. Error	Standardized Coefficients Beta	t	Sig.	Collinearity Statistics Tolerance	VIF
1	(Constant)	3.287	.081		40.592	.000		
	REGR factor score 1 for analysis 1	-.287	.506	-.215	-.567	.572	.026	38.812
	REGR factor score 2 for analysis 1	-.223	.093	-.168	-2.409	.017	.767	1.304
	REGR factor score 3 for analysis 1	.658	.354	.494	1.859	.065	.053	19.004
	REGR factor score 4 for analysis 1	-.138	.590	-.104	-.234	.815	.019	52.825
	REGR factor score 5 for analysis 1	-.186	.265	-.139	-.701	.485	.094	10.674
	REGR factor score 6 for analysis 1	-.157	.366	-.118	-.430	.668	.049	20.328
	REGR factor score 7 for analysis 1	.157	.216	.118	.725	.470	.141	7.091
	REGR factor score 8 for analysis 1	.926	1.025	.694	.903	.368	.006	159.354
	REGR factor score 9 for analysis 1	4801515.248	5581939.224	.911	.860	.391	.003	302.391

a. Dependent Variable: Log-transformed sales

On output of coefficients we see on Column Collinearity Statistics sub column all VIF value 1<5. This means problem collinearity has been overcome. The Condition Index Values show 1 <15 for all predictors. This have mean that the problem collinearity has been overcome. The last step is that we will run the linear regression with method using **stepwise**.

Collinearity Diagnostics^a

Model	Dimension	Eigenvalue	Condition Index	Variance Proportions (Constant)	REGR factor score 1 for analysis 1	REGR factor score 2 for analysis 1	REGR factor score 3 for analysis 1	REGR factor score 4 for analysis 1	REGR factor score 5 for analysis 1	REGR factor score 6 for analysis 1	REGR factor score 7 for analysis 1	REGR factor score 8 for analysis 1	REGR factor score 9 for analysis 1
1	1	1.998	1.000	.00	.00	.00	.00	.00	.00	.00	.00	.00	.00
	2	1.000	1.414	.00	.00	.19	.03	.00	.00	.00	.00	.00	.00
	3	1.000	1.414	.00	.00	.51	.01	.00	.00	.00	.00	.00	.00
	4	1.000	1.414	.00	.02	.03	.00	.01	.00	.00	.00	.00	.00
	5	1.000	1.414	.00	.00	.02	.00	.04	.00	.00	.00	.00	.00
	6	1.000	1.414	.59	.00	.00	.00	.00	.01	.01	.00	.00	.00
	7	1.000	1.414	.27	.00	.01	.00	.01	.03	.01	.00	.00	.00
	8	1.000	1.414	.13	.00	.00	.00	.00	.00	.12	.00	.00	.00
	9	1.000	1.414	.00	.00	.01	.00	.01	.03	.00	.00	.00	.00
	10	.002	34.750	.00	.98	.23	.95	.98	.91	.95	.86	1.00	1.00

a. Dependent Variable: Log-transformed sales

Do it the analysis using the following steps:
- **Analyze > Rgression > Linear**

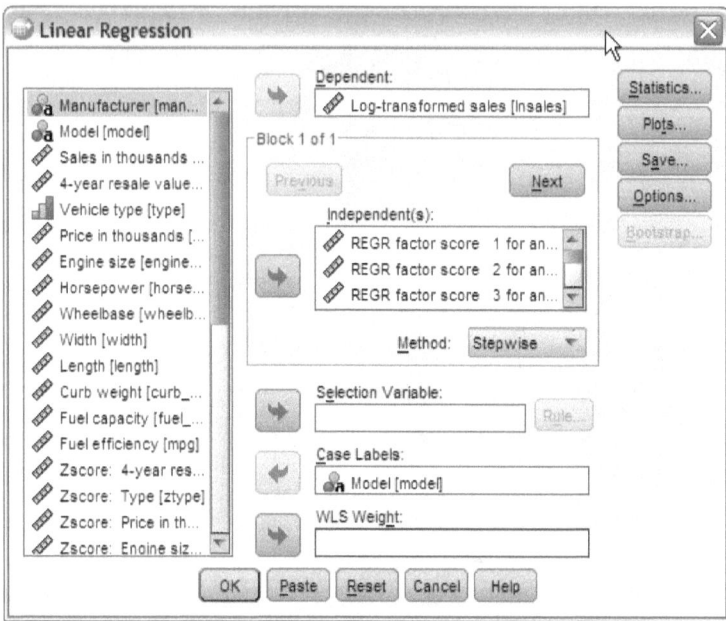

- Move the variable of **Log-transformed sales [Insales]** in **Dependent** column.
- Move all variables using the prefixed **REGR factor score** to **Independent (s)** column
- Move the **Model** variable **[model]** to **Case Labels** column
- choose **Method: Stepwise**
- choose **Statistics**

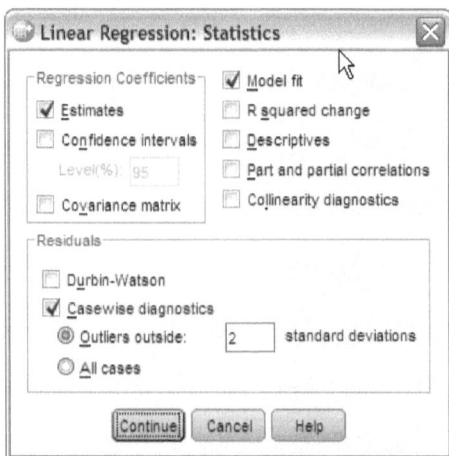

- On **Regression Coefficients** check on selection **Estimates** and **Fit Model**. On **Residuals** check **Casewise diagnostics** and choose **Outliers**

outside with enter the value of 2 for column **standard deviation** > **Continue**.

- choose **Plots**

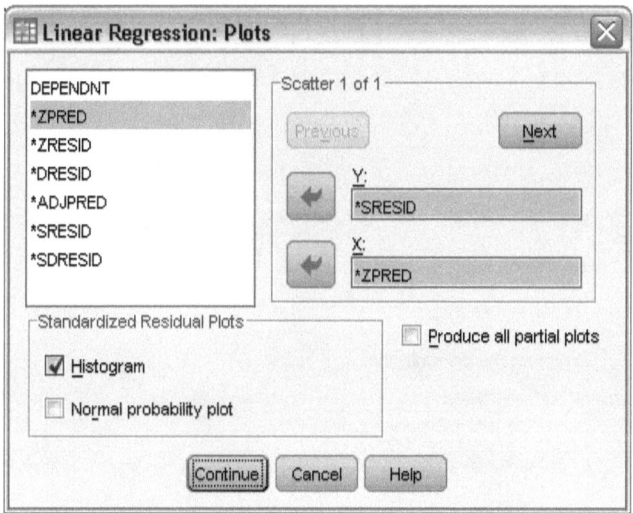

- Move ***ZPRED** to **X** and *** SDRESID** to **Y**. On selection **Standardized Residual Plots** check **Histogram** > **Continue**
- Select **Save**

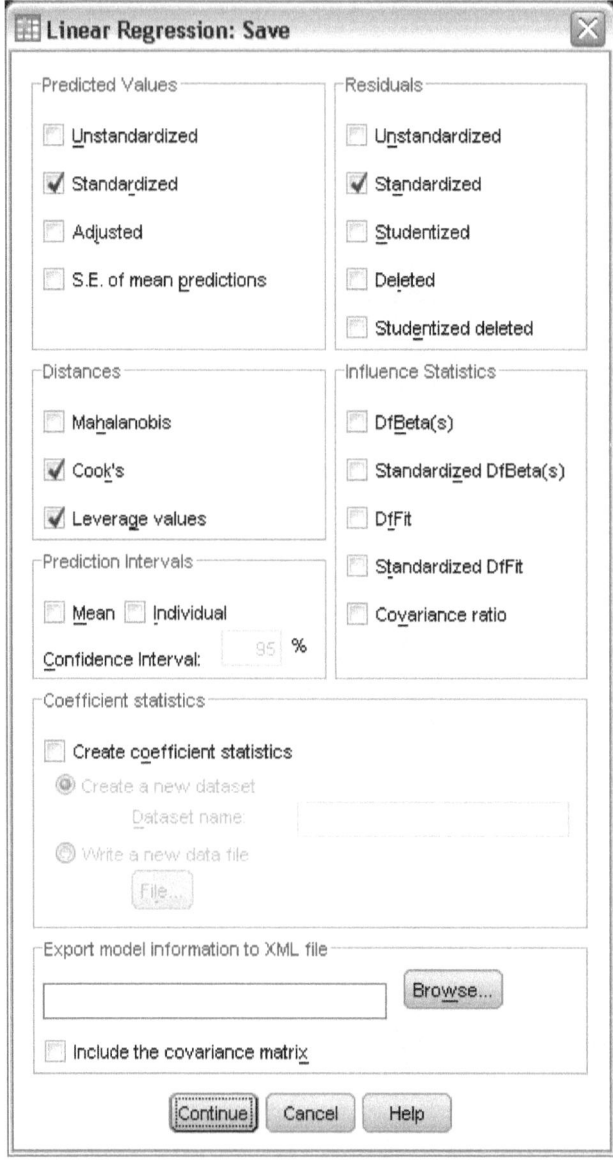

- On selection of **Predicted Values** check **standardized** . On selection **Residuals** also check **standardized**.
  - On selection **Distance**, check **Cook's** and **Leverage Values > Continue**
  - Click **OK**

Results and interpretation as below this:

Model Summary[e]

Model	R	R Square	Adjusted R Square	Std. Error of the Estimate
1	.537[a]	.289	.284	1.12828
2	.603[b]	.363	.355	1.07095
3	.661[c]	.437	.425	1.01052
4	.675[d]	.456	.441	.99649

a. Predictors: (Constant), REGR factor score 1 for analysis 1

b. Predictors: (Constant), REGR factor score 1 for analysis 1, REGR factor score 4 for analysis 1

c. Predictors: (Constant), REGR factor score 1 for analysis 1, REGR factor score 4 for analysis 1, REGR factor score 3 for analysis 1

d. Predictors: (Constant), REGR factor score 1 for analysis 1, REGR factor score 4 for analysis 1, REGR factor score 3 for analysis 1, REGR factor score 2 for analysis 1

e. Dependent Variable: Log-transformed sales

First Part: Summary Model

On part this we are given a selection with new models as many as 5 models. We can choose the best model with the highest R^2 value.

Second Part: ANOVA Table

ANOVA[a]

Model		Sum of Squares	df	Mean Square	F	Sig.
1	Regression	77.430	1	77.430	60.824	.000[b]
	Residual	190.953	150	1.273		
	Total	268.383	151			
2	Regression	97.489	2	48.745	42.500	.000[c]
	Residual	170.894	149	1.147		
	Total	268.383	151			
3	Regression	117.253	3	39.084	38.275	.000[d]
	Residual	151.130	148	1.021		
	Total	268.383	151			
4	Regression	122.412	4	30.603	30.819	.000[e]
	Residual	145.971	147	.993		
	Total	268.383	151			

a. Dependent Variable: Log-transformed sales

b. Predictors: (Constant), REGR factor score 1 for analysis 1

c. Predictors: (Constant), REGR factor score 1 for analysis 1, REGR factor score 4 for analysis 1

d. Predictors: (Constant), REGR factor score 1 for analysis 1, REGR factor score 4 for analysis 1, REGR factor score 3 for analysis 1

e. Predictors: (Constant), REGR factor score 1 for analysis 1, REGR factor score 4 for analysis 1, REGR factor score 3 for analysis 1, REGR factor score 2 for analysis 1

In the ANOVA table all the models are significant and correct because have significance value as much as 0. 00 < 0.05. Meaning that: there is a linear relationship between variables of *sales in thousands* with *fuel efficiency, length, price in thousand, vehicle type, width, engine size, wheel base, curb weight*, and *horse power*.

Data Analysis and Its Interpretation: Application in Marketing Research

Third Part: Regression Coeffecients

Coefficients[a]

Model		Unstandardized Coefficients		Standardized Coefficients	t	Sig.	Collinearity Statistics	
		B	Std. Error	Beta			Tolerance	VIF
1	(Constant)	3.289	.092		35.939	.000		
	REGR factor score 1 for analysis 1	-.716	.092	-.537	-7.799	.000	1.000	1.000
2	(Constant)	3.289	.087		37.863	.000		
	REGR factor score 1 for analysis 1	-.716	.087	-.537	-8.216	.000	1.000	1.000
	REGR factor score 4 for analysis 1	.364	.087	.273	4.182	.000	1.000	1.000
3	(Constant)	3.289	.082		40.127	.000		
	REGR factor score 1 for analysis 1	-.716	.082	-.537	-8.708	.000	1.000	1.000
	REGR factor score 4 for analysis 1	.364	.082	.273	4.432	.000	1.000	1.000
	REGR factor score 3 for analysis 1	.362	.082	.271	4.399	.000	1.000	1.000
4	(Constant)	3.289	.081		40.692	.000		
	REGR factor score 1 for analysis 1	-.716	.081	-.537	-8.830	.000	1.000	1.000
	REGR factor score 4 for analysis 1	.364	.081	.273	4.495	.000	1.000	1.000
	REGR factor score 3 for analysis 1	.362	.081	.271	4.461	.000	1.000	1.000
	REGR factor score 2 for analysis 1	-.185	.081	-.139	-2.279	.024	1.000	1.000

a. Dependent Variable: Log-transformed sales

The above table shows that regression coefficients are all significant because the values of significance are less than 0.05.

Fourth Part: Histogram and Scatterplot

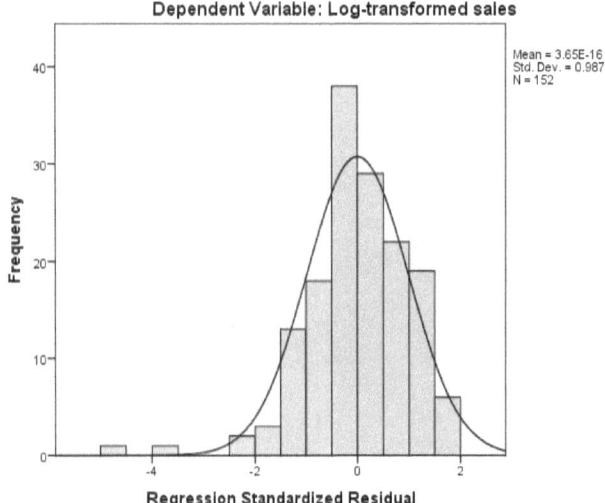

The graph above show normality of data already we analysis in the regression model this .

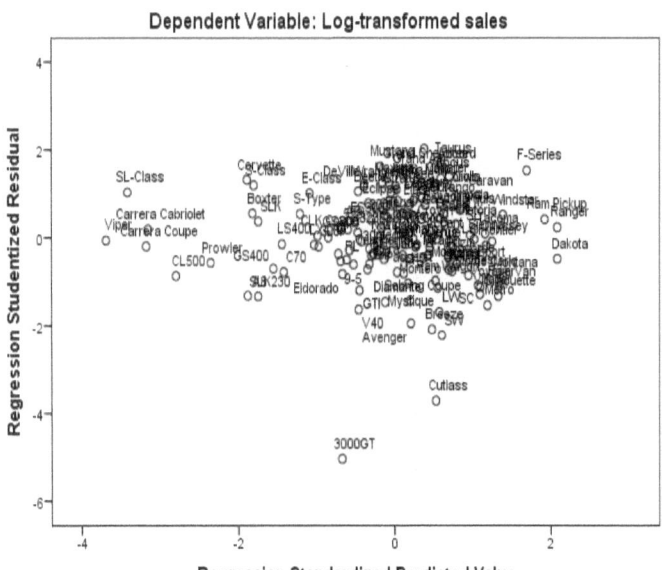

On picture above the scatter plot above shows that the data that does not form certain patterns. This means that the regression model is correct because there is no heteroscedasticity.

8.8.3 Conclusion

From the above analysis we could make conclusion as the following :

- The regression model that we made is already feasible and right .
- There is linear relationship between variables of *sales in thousands* with *fuel efficiency, length, price in thousand, vehicle type, width, engine size, wheel base, curb weight* , and *horse power* . In other words, variable *fuel efficiency, length, price in thousand, vehicle type, width, engine size, wheel base, curb weight* , and *horse power* take effect on signify to variable *sales in thousands* .

CHAPTER 9

DIRECT MARKETING: PROCEDURES USING CONSUMERS' DATA BASE

9.1 RFM Analysis

RFM analysis is a technique for identifying existing consumers who most often respond to new product / service offerings offered to them. This technique is commonly used in direct marketing based on the simple theory that says: 1) The most important factor in identifying customers who are likely to respond to the new offer is recency. The point is that consumers who buy recently will tend to buy again compared to consumers who bought in the past; 2) The second most important factor is the frequency. The point is that consumers who more often make purchases in the past will likely respond more when compared to consumers who rarely make purchases in the past; 3) The third most important factor is the total amount of money spent (monetary). That is, consumers who have shopped in more numbers will tend to respond more when compared with consumers who spend less money.

How does RFM analysis work ? How it works is as follows:

- Consumers are charged the recency score based on the most recent purchase date interval since the last purchase. This value is based on the simple ranking of the novelty value into a small number of categories. As a matter of fact, if we use five categories, then consumers who date the most recent purchases will get a ranking of 5, otherwise the latter will get a value of 1.

- In the same way as above, consumers are subject to higher frequency rankings for those who make frequent purchases and are small for those who rarely make a purchase. For example, using rank 1 to 5, then consumers who frequent the frequency of spending more often rated the rank 5 while the least one is given a value of 1

- Third, consumers in rankings are based on monetary value , for consumers who spend the most money will get a ranking of 5; while the least will get 1

- The end result is four values for each consumer: the recency, frequency, monetary, and combined RFM values, which are simply three values each merged into a single value. The best consumers will get a combined value of RFM. For example, in the ranking of five categories, there is a total value of 125 and a top score of 555.

Data Analysis and Its Interpretation: Application in Marketing Research

What data form can be analyzed using RFM technique? Data are:

- If data represents transaction (where each row represents a single transaction, and there are many transactions for each consumer), then use RFM from Transactions.

- If data represent consumers with summary information for all transactions (with columns containing total number issued, total number of transactions, and latest transaction date), then use RFM from Customer Data.

9.2 RFM analysis

To perform RFM procedures in IBM SPSS the steps are as follows:

First : Prepare the Data and do the analysis as below:
- Open File: rfm_transaction.sav with the number of ID variables, Product Line, Product Number, Date, and Amount and amount of data 4906
- Select the Direct Marketing menu
- Choose Techniques
- Select Help Identify my Best Contact (RFM Analysis), press Continue
- In the RFM Data Format option, select Transaction Data, tap Continue
- Move the Purchase Date (Date) variable to the Transaction Date column, the Purchase Amount variable (Amount) to the Transaction Amount, and the Customer ID (ID) to the Customer Identifiers
- Click OK
  The result will be as below:

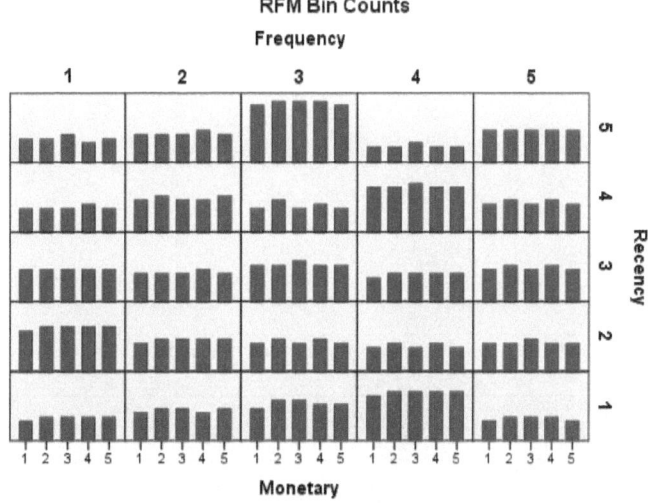

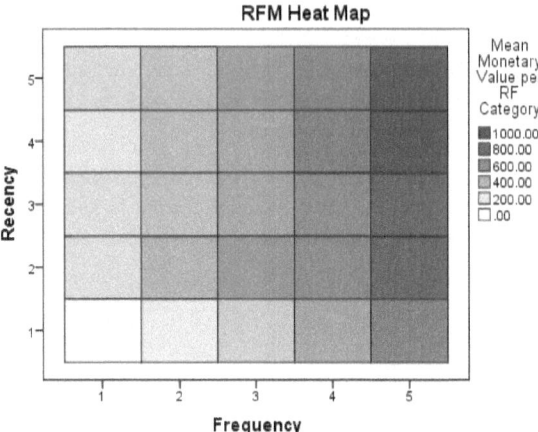

Second : Interpretation of results

Interpretation of output is done by looking at new variables in the new Data Set analysis results, namely:

1. ID (Customer ID)
2. Date Most Recent
3. Transaction Count
4. Amount (Number of Transactions)
5. Recency Score
6. Frequency Score
7. Monetary Score
8. RFM Score

Example of interpretation: Let us take consumer 1 and 2.

ID	Date Most Recent	Transaction Count	Amount	Recency Score	Frequency Score	Monetary Score	RFM Score
1	04-Sep-2006	5	485	4	3	4	434
2	10-Nov-2005	4	350	2	2	2	222

- Subscriber with ID number 1, making transaction on September 4, 2006. Number of transactions 5 times with total purchase of 485. Related to get recency score of 4, frequency score of 3 and monetary score of 4 from maximum value of 5. While value of RFM combined amounting to 434 from the highest score of 555.

- Subscriber with ID number 2, making transaction on November 10, 2005. Number of transactions 4 times with total purchase of 350. Related to get recency score of 2, frequency score of 2 and monetary score of 2 from maximum value of 5. While value of RFM combined amounting to 222 from the highest score of 555.

From both data above we know that customer with ID number 1 is more potential customer compared with customer with ID number 2. Thus the RFM theory applied to both cases, namely recency, frequency and monetary can be used as the basis of prediction which customer will be more responsive to new offerings from the company.

References

Anderson, Sweeny & William.(2011). *Statistics for Business and Economics*. South Western: Cengage Learning

Brown, James Dean. (1988). *Understanding Research in Second Language Learning*. Cambridge:Cambridge University Press

Cramer, Duncan and Howitt, Dennis. (2006) *The Sage Dictionary of Statistics*. London: Sage Publication.

Field, Andy. (2006) *Discovering Statistics Using SPSS*. Second Edition. London: SAGE Publications Ltd

Gujarati, D. N. & Porter, D.C.(2009). *Basic Econometrics*. (5th ed.)New York: MacGraw Hill

Gujarati, Damodar. N. (2006). *Essentials of Econometrics*. New York: MacGraw Hill

IBM SPSS Brief Guide Version 25

IBM SPSS Statistics Core Users' Guide Version 25

IBM Statistics Information Center.
http://publib.boulder.ibm.com/infocenter/spssstat/v20r0m0/index.jsp

Johnstom, Ian. (2004). *A Normal Distribution and The Normal Curve*. Malaspina University-College

Levin, R.I & Rubin, S.D.(1998). *Statistics For Management*. (7th ed.). New Jersey:

Prentice – Hall

Sarwono, J.(2012) SPSS 20 Aplikasi untuk Riset Eksperimental. Jakarta: Penerbit Elexmedia Komputindo

Sarwono, J.(2013) 12 Jurus Ampuh Regresi untuk Riset Skripsi. Jakarta: Penerbit Elexmedia Komputindo

Tuckman (1978) Conducting Educational Research. New York: HBC Publishers

Zikmund, W. G. (1997). *Exploring Marketing Research*. Florida: The Dryden Press.

ABOUT THE AUTHOR

Dewi Indriani Jusuf is the Rector at International Women University Bandung Indonesia. She also teaches at Post Graduate Program in several universities in Bandung. Besides, she is also a women organizational activtist in Indonesia at present. She can be contacted at dewijusuf@iwu.ac.id

Jonathan Sarwono currently is the Director of Quality Assurance in International Women University Bandung, Indonesia. He is also a lecturer in some universities in Bandung and Jakarta. He can be contacted at jsarwono007@gmail.com.

www.ingramcontent.com/pod-product-compliance
Lightning Source LLC
Chambersburg PA
CBHW031622210526
45464CB00004B/1706